Lecture Notes in Computer Science   **11666**

More information about this series at http://www.springer.com/series/7410

Marco Baldi · Edoardo Persichetti ·
Paolo Santini (Eds.)

Code-Based Cryptography

7th International Workshop, CBC 2019
Darmstadt, Germany, May 18–19, 2019
Revised Selected Papers

 Springer

Editors
Marco Baldi [ID]
Marche Polytechnic University
Ancona, Italy

Edoardo Persichetti [ID]
Florida Atlantic University
Boca Raton, FL, USA

Paolo Santini [ID]
Marche Polytechnic University
Ancona, Italy

ISSN 0302-9743   ISSN 1611-3349   (electronic)
Lecture Notes in Computer Science
ISBN 978-3-030-25921-1   ISBN 978-3-030-25922-8   (eBook)
https://doi.org/10.1007/978-3-030-25922-8

LNCS Sublibrary: SL4 – Security and Cryptology

This Springer imprint is published by the registered company Springer Nature Switzerland AG
The registered company address is: Gewerbestrasse 11, 6330 Cham, Switzerland

Preface

Code-based cryptography is the branch of cryptography that relies on the hardness of some decoding problems that are recognized not to be significantly affected by quantum computers.

It follows that code-based cryptography is one of the fields of major interest in the context of post-quantum cryptography. The importance of this field has been steadily increasing in the last few years, also thanks to the launch of important activities such as the NIST Post-Quantum Cryptography standardization call.

The Code-Based Cryptography Workshop (CBC) was born in 2009 thanks to the effort of a selected group of researchers, located mainly between France and the Netherlands, who sought to bring together the community and promote the study of this field. Initially organized as a pure workshop-like forum on a biennial basis, it has since grown and attracted increasing interest, crossing the Atlantic for the first time last year in its 6th edition, held in the United States of America. The aim of this 7th edition (CBC 2019), held in Darmstadt, Germany during May 18–19, 2019, was to continue gathering researchers working in the field and promote interest, by providing a forum to discuss cutting-edge research, as well as share visions for future advances.

Besides contributed talks, the chance to submit full papers for the inclusion in these proceedings was introduced in CBC 2019. The workshop attracted 14 submissions along this line, of which 8 were selected by the Program Committee through careful peer-review, and are collected in this book.

These contributions are divided into two groups: the first four papers deal with the design of code-based cryptosystems, while the following four papers are on cryptanalysis of code-based cryptosystems. As such, the works presented in this book provide a synthetic yet significant overview of the state of the art of code-based cryptography, laying out the groundwork for future developments.

The workshop was enriched with two invited presentations by internationally recognized researchers, Daniel J. Bernstein and Ray Perlner, to whom we are grateful. Furthermore, the program included eight contributed talks, presenting recent research and works in progress, as well as working and discussion sessions.

We would like to thank the Program Committee members for their strenuous work on the submissions, and the external reviewers for their valuable contribution. We are also very grateful to our sponsors, Darkmatter LLC and Oak Ridge Associated Universities (ORAU), for their generous support. Finally, we thank the Organizing Committee of Eurocrypt 2019 for allowing us to host this event in co-location with their conference, which greatly contributed to its success.

May 2019

Marco Baldi
Edoardo Persichetti
Paolo Santini

Organization

Program Committee

Marco Baldi	Marche Polytechnic University, Italy
Paulo Barreto	University of Washington, USA
Pierre-Louis Cayrel	Laboratoire Hubert Curien, France
Franco Chiaraluce	Marche Polytechnic University, Italy
Marche Tung Chou	Osaka University, Japan
Taraneh Eghlidos	Sharif University of Technology, Iran
Philippe Gaborit	University of Limoges, France
Tim Güneysu	Ruhr-Universität Bochum, DFKI, Germany
Thomas Johansson	Lund University, Sweden
Irene Marquez-Corbella	Universidad de La Laguna, Spain
Ayoub Otmani	Université de Rouen Normandie, UR, LITIS, France
Edoardo Persichetti	Florida Atlantic University, USA
Joachim Rosenthal	University of Zurich, Switzerland
Paolo Santini	Marche Polytechnic University, Italy
Nicolas Sendrier	Inria, France
Jean-Pierre Tillich	Inria, France
Øyvind Ytrehus	University of Bergen, Norway

Additional Reviewers

Ahmad Khan, Junaid
Bagheri, Mina
Battaglioni, Massimo
Khathuria, Karan
von Maurich, Ingo
Weger, Violetta

Contents

Quantum Resistant Public Key Encryption Scheme HermitianRLCE

Gretchen L. Matthews[1] and Yongge Wang[2]([✉])

[1] Virginia Polytechnic Institute and State University, Blacksburg, VA 24061, USA
gmatthews@vt.edu
[2] UNC Charlotte, 9201 University City Blvd., Charlotte, NC 28223, USA
yonwang@uncc.edu

Abstract. Recently, Wang (2017) introduced a random linear code based quantum resistant public key encryption scheme RLCE which is a variant of McEliece encryption scheme. Wang (2017) analyzed an instantiation of RLCE scheme using Generalized Reed-Solomon codes. In this paper, we introduce and analyze Hermitian code based RLCE schemes HermitianRLCE. Based on our security analysis, we provide Hermitian-RLCE parameters at the 128, 192, and 256 bits security level. These parameters show that HermitianRLCE has much smaller public keys than GRS-RLCE.

Keywords: Random linear codes · McEliece encryption scheme · Linear code based encryption scheme

1 Introduction

Since McEliece encryption scheme [8] was introduced more than thirty years ago, it has withstood many attacks and still remains unbroken for general cases. It has been considered as one of the candidates for post-quantum cryptography since it is immune to existing quantum computer algorithm attacks. The original McEliece cryptography system is based on binary Goppa codes. Several variants have been introduced to replace Goppa codes in the McEliece encryption scheme though most of them have been broken. Up to the writing of this paper, secure McEliece encryption schemes include MDPC/LDPC code based McEliece encryption schemes [1,9], Wang's RLCE [12,13], and the original binary Goppa code based McEliece encryption scheme. The advantage of the RLCE encryption scheme is that its security does not depend on any specific structure of underlying linear codes, instead its security is believed to depend on the NP-hardness of decoding random linear codes.

The RLCE scheme [12,13] could be used as a template to design encryption schemes based on any linear codes. Wang [12,13] analyzed Generalized Reed-Solomon code based RLCE security. This paper proposes a Hermitian code based

The first author is supported by NSF-DMS 1855136.
The second author is supported by Qatar Foundation Grant NPRP8-2158-1-423.

M. Baldi et al. (Eds.): CBC 2019, LNCS 11666, pp. 1–10, 2019.
https://doi.org/10.1007/978-3-030-25922-8_1

RLCE scheme. It is shown that Hermitian code based RLCE scheme has smaller key sizes compared with Generalized Reed-Solomon code based RLCE schemes. For example, for the AES 128, 192, and 256 security levels, the GRS-RLCE schemes have public keys of size 183 KB, 440 KB, and 1203 KB respectively. For HermitianRLCE schemes, the corresponding public keys are of the size 103 KB, 198 KB, 313 KB respectively. It should be noted that several authors have tried to design algebraic-geometric code based McEliece encryption scheme (see, e.g., [6]). However, most of these algebraic-geometric code based McEliece encryption schemes have been broken (see, e.g., [3]). Hermitian RLCE provides an alternative approach which combines the algebraic geometric construction based on Hermitian curves (and more generally, the extended norm-trace curves) with that of a random linear code.

Unless specified otherwise, bold face letters such as $\mathbf{a}, \mathbf{b}, \mathbf{e}, \mathbf{f}, \mathbf{g}$ are used to denote row or column vectors over $\mathbb{F}_q$. It should be clear from the context whether a specific bold face letter represents a row vector or a column vector.

2   McEliece and RLCE Encryption Schemes

For given parameters n, k and t, the McEliece scheme [8] chooses an $[n, k, 2t+1]$ linear Goppa code $\mathcal{C}$. Let G_s be the $k \times n$ generator matrix for the code $\mathcal{C}$. Select a random dense $k \times k$ non-singular matrix S and a random $n \times n$ permutation matrix P. Then the public key is $G = SG_sP$ and the private key is G_s. The following is a description of encryption and decryption processes.

Mc.Enc$(G, \mathbf{m}, \mathbf{e})$. For a message $\mathbf{m} \in \{0, 1\}^k$, choose a random vector $\mathbf{e} \in \{0, 1\}^n$ of weight t and compute the cipher text $\mathbf{c} = \mathbf{m}G + \mathbf{e}$

Mc.Dec$(S, G_s, P, \mathbf{c})$. For a received ciphertext $\mathbf{c}$, first compute $\mathbf{c}' = \mathbf{c}P^{-1} = \mathbf{m}SG$. Next use an error-correction algorithm to recover $\mathbf{m}' = \mathbf{m}S$ and finally compute the message $\mathbf{m}$ as $\mathbf{m} = \mathbf{m}'S^{-1}$.

The protocol for the RLCE Encryption scheme by Wang [12] consists of the following three processes: RLCE.KeySetup, RLCE.Enc, and RLCE.Dec. Specifically the revised RLCE scheme proceeds as follows.

RLCE.KeySetup(n, k, d, t, w). Let n, $k, d, t > 0$, and $w \in \{1, \cdots, n\}$ be given parameters such that $n - k + 1 \geq d \geq 2t + 1$. Let G_s be a $k \times n$ generator matrix for an $[n, k, d]$ linear code $\mathcal{C}$ such that there is an efficient decoding algorithm to correct at least t errors for this linear code given by G_s. Let P_1 be a randomly chosen $n \times n$ permutation matrix and $G_sP_1 = [\mathbf{g}_0, \cdots, \mathbf{g}_{n-1}]$.

1. Let $\mathbf{r}_0, \mathbf{r}_1, \cdots, \mathbf{r}_{w-1} \in \mathbb{F}_q^k$ be column vectors drawn uniformly at random and let

$$G_1 = [\mathbf{g}_0, \cdots, \mathbf{g}_{n-w}, \mathbf{r}_0, \cdots, \mathbf{g}_{n-1}, \mathbf{r}_{w-1}] \tag{1}$$

be the $k \times (n + w)$ matrix obtained by inserting column vectors $\mathbf{r}_i$ into G_sP_1.

2. Let $A_0 = \begin{pmatrix} a_{0,00} & a_{0,01} \\ a_{0,10} & a_{0,11} \end{pmatrix}, \cdots, A_{w-1} = \begin{pmatrix} a_{w-1,00} & a_{w-1,01} \\ a_{w-1,10} & a_{w-1,11} \end{pmatrix} \in \mathbb{F}_q^{2\times2}$ be non-singular 2×2 matrices chosen uniformly at random such that $a_{i,00}a_{i,01}a_{i,10}a_{i,11} \neq 0$ for all $i = 0, \cdots, w-1$. Let $A = \text{diag}[1, \cdots, 1, A_0, \cdots, A_{w-1}]$ be an $(n+w) \times (n+w)$ non-singular matrix.

3. Let S be a random dense $k \times k$ non-singular matrix and P_2 be an $(n+w) \times (n+w)$ permutation matrix.

4. The public key is the $k \times (n+w)$ matrix $G = SG_1AP_2$ and the private key is (S, G_s, P_1, P_2, A).

RLCE.Enc$(G, \mathbf{m}, \mathbf{e})$. For a row vector message $\mathbf{m} \in \mathbb{F}_q^k$, choose a random row vector $\mathbf{e} = [e_0, \ldots, e_{n+w-1}] \in \mathbb{F}_q^{n+w}$ such that the Hamming weight of $\mathbf{e}$ is at most t. The cipher text is $\mathbf{c} = \mathbf{m}G + \mathbf{e}$.

RLCE.Dec$(S, G_s, P_1, P_2, A, \mathbf{c})$. For a received cipher text $\mathbf{c} = [c_0, \ldots, c_{n+w-1}]$, compute

$$\mathbf{c}P_2^{-1}A^{-1} = \mathbf{m}SG_1 + \mathbf{e}P_2^{-1}A^{-1} = [c_0', \ldots, c_{n+w-1}'].$$

Let $\mathbf{c}' = [c_0', c_1', \cdots, c_{n-w}', c_{n-w+2}', \cdots, c_{n+w-2}']$ be the row vector of length n selected from the length $n+w$ row vector $\mathbf{c}P_2^{-1}A^{-1}$. Then $\mathbf{c}'P_1^{-1} = \mathbf{m}SG_s + \mathbf{e}'$ for some error vector $\mathbf{e}' \in \mathbb{F}_q^n$ where the Hamming weight of $\mathbf{e}' \in \mathbb{F}_q^n$ is at most t. Using an efficient decoding algorithm, one can recover $\mathbf{m}SG_s$ from $\mathbf{c}'P_1^{-1}$. Let D be a $k \times k$ inverse matrix of SG_s' where G_s' is the first k columns of G_s. Then $\mathbf{m} = \mathbf{c}_1 D$ where $\mathbf{c}_1$ is the first k elements of $\mathbf{m}SG_s$. Finally, calculate the Hamming weight $wt = \text{wt}(\mathbf{c} - \mathbf{m}G)$. If $wt \leq t$ then output $\mathbf{m}$ as the decrypted plaintext. Otherwise, output error.

3 Hermitian Codes

Consider the curve X given by

$$y^{q^{r-1}} + y^{q^{r-2}} + \cdots + y^q + y = x^u \tag{2}$$

over the field $\mathbb{F}_{q^r}$ where $u|\frac{q^r-1}{q-1}$. Notice that when $u = \frac{q^r-1}{q-1}$, the Eq. (2) gives

$$Tr_{\mathbb{F}_{q^r}/\mathbb{F}_q} = N_{\mathbb{F}_{q^r}/\mathbb{F}_q}.$$

When $r = 2$ and $u = \frac{q^r-1}{q-1}$, the Eq. (2) gives

$$y^q + y = x^{q+1}$$

over $\mathbb{F}_{q^2}$ which is the defining equation of the Hermitian curve. In general, the genus of X is

$$g = \frac{(u-1)(q^{r-1}-1)}{2}$$

and X has at least

$$\bar{n} = q^{r-1} + u(q^r - q^{r-1})$$

Table 1. Parameters for Hermitian curves

q	r	u	$\bar{n}$	g
16	2	17	4096	120
4	4	5	1024	126
		17	3328	504
		85	16384	2646
2	8	3	512	127
		5	768	254
		15	2048	889
		17	2304	1016
		51	6656	3175
		85	11008	5334
		255	32768	16129

affine $\mathbb{F}_{q^r}$-rational points. We are interested in taking $q^r = 2^8$ which gives several options for r and q as shown in Table 1.

The Hermitian code over $\mathbb{F}_{q^2}$ is defined using the Hermitian curve $y^q + y = x^{q+1}$. To begin, fix $n \leq q^3$ and $2g + 1 < \alpha < n$. Select n distinct $\mathbb{F}_{q^2}$-rational affine points $P_1, \ldots, P_n$ on X. Thus, each P_i is of the form P_{ab} where $a, b \in \mathbb{F}_{q^2}$ and $b^q + b = a^{q+1}$. Let $D = P_1 + \cdots + P_n$, and set

$$\mathcal{B} := \{x^i y^j : i \geq 0, 0 \leq j \leq q - 1, iq + j(q - 1) \leq \alpha\};$$

one may note that $\mathcal{B}$ is a basis for the vector space $\mathcal{L}(\alpha P)$, where P denotes the point at infinity on X. In the Hermitian code $C(D, \alpha P)$, the message is a polynomial $f \in \text{Span}(\mathcal{B})$, and the codeword of the message polynomial f is the evaluations of f over the Hermitian curve. More precisely, $C = C(D, \alpha P)$ is the image of the evaluation map

$$ev : \mathcal{L}(\alpha P) \rightarrow \qquad \mathbb{F}_{q^2}^n$$
$$f \qquad \mapsto (f(P_1), \cdots, f(P_n)).$$

It is noted that for the Hermitian code $C(D, \alpha P)$ with $\alpha \geq 2g - 1$, the length is n, the dimension is $\alpha + 1 - g$, and the minimum distance satisfies $d \geq n - \alpha$. The exact minimum distances are known [14].

4   HermitianRLCE

HermitianRLCE is an RLCE encryption scheme with Hermitian code as the underlying code. Specifically, HermitianRLCE replaces the the generator matrix G_s utilized in RLCE.KeySetup(n, k, d, t, w) in Sect. 2 with a Hermitian code $k \times n$ generator matrix.

5   Security Analysis

In the following sections, we carry out heuristic security analyses on the revised RLCE scheme.

5.1   Classical and Quantum Information-Set Decoding

Information-set decoding (ISD) is one of the most important message recovery attacks on McEliece encryption schemes. The state-of-the-art ISD attack for non-binary McEliece scheme is the one presented in Peters [10], which is an improved version of Stern's algorithm [11]. For the RLCE encryption scheme, the ISD attack is based on the number of columns in the public key G instead of the number of columns in the private key G_s. The cost of ISD attack on an $[n, k, t; w]$-RLCE scheme is equivalent to the cost of ISD attack on an $[n+w, k; t]$-McEliece scheme.

For the naive ISD, one first uniformly selects k columns from the public key and checks whether it is invertible. If it is invertible, one multiplies the inverse with the corresponding ciphertext values in these coordinates that correspond to the k columns of the public key. If these coordinates contain no errors in the ciphertext, one recovers the plain text. To be conservative, we may assume that randomly selected k columns from the public key is invertible. For each $k \times k$ matrix inversion, Strassen algorithm takes $O(k^{2.807})$ field operations (though Coppersmith-Winograd algorithm takes $O(k^{2.376})$ field operations in theory, it may not be practical for the matrices involved in RLCE encryption schemes). In a summary, the naive information-set decoding algorithm takes approximately $2^{\kappa'_c}$ steps to find k-error free coordinates where, by Sterling's approximation,

$$
\begin{aligned}
\kappa'_c &= \log_2 \left(\frac{\binom{n+w}{k} (k^{2.807} + k^2)}{\binom{n+w-t}{k}} \right) \\
&\simeq (n+w) I \left(\frac{k}{n+w} \right) - (n+w-t) I \left(\frac{k}{n+w-t} \right) + \log_2 \left(k^{2.807} + k^2 \right)
\end{aligned}
\tag{3}
$$

and $I(x) = -x \log_2(x) - (1 - x) \log_2(1 - x)$ is the binary entropy of x. There are several improved ISD algorithms in the literature. These improved ISD algorithms allow a small number of error positions within the selected k ciphertext values or select $k + \delta$ columns of the public key matrix for a small number $\delta > 0$ or both.

An HermitianRLCE scheme is said to have quantum security level κ_q if the expected running time (or circuit depth) to decrypt a HermitianRLCE ciphertext using Grover's algorithm based ISD is 2^{κ_q}. For a function $f : \{0, 1\}^l \rightarrow \{0, 1\}$ with the property that there is an $x_0 \in \{0, 1\}^l$ such that $f(x_0) = 1$ and $f(x) = 0$ for all $x \neq x_0$, Grover's algorithm finds the value x_0 using $\frac{\pi}{4}\sqrt{2^l}$ Grover iterations and $O(l)$ qubits. Specifically, Grover's algorithm converts the function f to a reversible circuit C_f and calculates

$$
|x\rangle \xrightarrow{C_f} (-1)^{f(x)} |x\rangle
$$

in each of the Grover iterations, where $|x\rangle$ is an l-qubit register. Thus the total steps for Grover's algorithm is bounded by $\frac{\pi |C_f|}{4}\sqrt{2^l}$.

For the HermitianRLCE scheme, the quantum ISD attack first uniformly selects k columns from the public key and checks whether it is invertible. If it is invertible, one multiplies the inverse with the ciphertext. If these coordinates contain no errors in the ciphertext, one recovers the plain text. Though Grover's algorithm requires that the function f evaluate to 1 on only one of the inputs, there are several approaches (see, e.g., Grassl et al. [5]) to cope with cases that f evaluates to 1 on multiple inputs.

For randomly selected k columns from a RLCE encryption scheme public key, the probability that the ciphertext contains no errors in these positions is $\frac{\binom{n+w-t}{k}}{\binom{n+w}{k}}$. Thus the quantum ISD algorithm requires $\sqrt{\binom{n+w}{k}/\binom{n+w-t}{k}}$ Grover iterations. For each Grover iteration, the function f needs to carry out the following computations:

1. Compute the inverse of a $k \times k$ sub-matrix G_{sub} of the public key and multiply it with the corresponding entries within the ciphertext. This takes $O\left(k^{2.807} + k^2\right)$ field operations if Strassen algorithm is used.
2. Check that the selected k positions contain no errors in the ciphertext. This can be done with one of the following methods:
   (a) Multiply the recovered message with the public key and compare the differences from the ciphertext. This takes $O((n + w)k)$ field operations.
   (b) Use the redundancy within message padding scheme to determine whether the recovered message has the correct padding information. The cost for this operation depends on the padding scheme.

It is expensive for circuits to use look-up tables for field multiplications. Using Karatsuba algorithm, Kepley and Steinwandt [7] constructed a field element multiplication circuit with gate counts of $7 \cdot (\log_2 q^2)^{1.585}$. In a summary, the above function f for the HermitianRLCE quantum ISD algorithm could be evaluated using a reversible circuit C_f with $O\left(7\left((n + w)k + k^{2.807} + k^2\right)(\log_2 q^2)^{1.585}\right)$ gates. To be conservative, we may assume that a randomly selected k-columns sub-matrix from the public key is invertible. Thus Grover's quantum algorithm requires approximately

$$7\left((n + w)k + k^{2.807} + k^2\right)(\log_2 q^2)^{1.585}\sqrt{\frac{\binom{n+w}{k}}{\binom{n+w-t}{k}}} \qquad (4)$$

steps for the simple ISD algorithm against HermitianRLCE encryption scheme.

5.2 Schur Product Attacks on Algebraic Geometric Codes

Couvreur, Marquez-Corbella, and Pellikaan [4] introdcued a Schur product based attack on algebraic geometry codes based McEliece encryption schemes. Their attack can decrypt any encrypted message in $O(n^3)$ operations after computing an Error Correcting Pair in $O(n^4)$ operations. Specifically, their attack works for high genus algebraic geometry codes. In this section, we show how to choose parameters for HermitianRLCE scheme to avoid the attacks in [4].

For two codes C_1 and C_2 of length n, the star product code $C_1 * C_2$ is the vector space spanned by $\mathbf{a} * \mathbf{b}$ for all pairs $(\mathbf{a}, \mathbf{b}) \in C_1 \times C_2$ where $\mathbf{a} * \mathbf{b} = [a_0 b_0, a_1 b_1, \cdots, a_{n-1} b_{n-1}]$. For the square code $C^2 = C * C$, we have $\dim C^2 \leq \min \left\{ n, \binom{\dim C + 1}{2} \right\}$. For an $[n, k]$ Hermitian code C with $2g < \alpha < \frac{n+1}{2}$, it follows from [4, Corollary 6] that $\dim C^2 = 2k + g - 1$. The following is a brief proof on this fact. Note that $g = \frac{q(q-1)}{2}$. Fix $C = C(\alpha P)$. We claim $C^2 = C(2\alpha P)$. Suppose $W \in C^2$. Then $W = (f(P_1)h(P_1), \cdots, f(P_n)h(P_n)) = (fh(P_1), \cdots, fh(P_n))$ for some $f, h \in \mathcal{L}(\alpha P)$. But $fh \in \mathcal{L}(2\alpha P)$ since $(fh) = (f) + (h) \geq -\alpha P - \alpha P = -2\alpha P$. Thus we have $W \in C(2\alpha P)$. An argument in [4] could be used to show that $C(2\alpha P)$.

Let G be the public key for an (n, k, d, t, w) HermitianRLCE encryption scheme based on a Hermitian code. Let C be the code generated by the rows of G. Let $\mathcal{D}_1$ be the code with a generator matrix D_1 obtained from G by replacing the randomized $2w$ columns with all-zero columns and let $\mathcal{D}_2$ be the code with a generator matrix D_2 obtained from G by replacing the $n - w$ non-randomized columns with zero columns. Since $C \subset \mathcal{D}_1 + \mathcal{D}_2$ and the pair $(\mathcal{D}_1, \mathcal{D}_2)$ is an orthogonal pair, we have $C^2 \subset \mathcal{D}_1^2 + \mathcal{D}_2^2$. It follows that

$$2k + g - 1 \leq \dim C^2 \leq \min\{2k + g - 1, n - w\} + 2w \tag{5}$$

where we assume that $2w \leq k^2$. In the following discussion, we assume that *the $2w$ randomized columns in $\mathcal{D}_2$ behave like random columns in the attacks of* [4]. In all of our selected HermitianRLCE parameters, we have $2k + g - 1 > n - w$. Thus $\dim C^2 = \mathcal{D}_1^2 + \mathcal{D}_2^2 = n - w + \mathcal{D}_2^2 = n + w$. Furthermore, for any code C' of length n' that is obtained from C using code puncturing and code shortening, we have $\dim C'^2 = n'$. Thus the techniques in [4] could not be used to recover any non-randomized columns in $\mathcal{D}_1$.

As we have mentioned in the preceding paragraph, our selected parameters satisfies the condition $2k + g - 1 > n - w$. Thus plain filtration attacks will not identify the randomized columns. However, one may select $w' < w$ columns from the public key and shorten these w' columns. A similar analysis as in Couvreur, Lequesne, and Tillich [2] shows that if these w' columns are the added random columns, then the resulting code is a $(k - w') \times (n - w')$ HermitianRLCE code with $w - w'$ added random columns. In order for one to verify that w' columns are added random columns, one needs to observe that

$$2(k - w') + g - 1 + w'^2 < \min\{(k - w')^2, n - w'\} \tag{6}$$

In our parameter selection, we make sure that for all $w' < w$, the inequality (6) does not hold.

6 Recommended Parameters

In this section, we propose parameters for HermitianRLCE schemes with equivalent security levels of AES-128, AES-192, and AES-256. If we take the code

Table 2. Parameters for HermitianRLCE scheme with $q = 16$, $r = 2$, and $u = 17$

ID	κ_c	κ_q	n	k	t	w	cipher bytes	sk	pk bytes	GRS-RLCE pk bytes
0	128	84	635	280	118	22	635		105560	188001
1	192	118	870	421	165	33	870		202922	450761
2	256	148	1090	531	220	45	1090		320724	1232001

$C(D, \alpha P)$ where $D = P_1 + \cdots + P_n$ and $q(q-1) < \alpha < n$, then we have $k = \alpha + 1 - g$ and $d \geq n - \alpha$. That is, the Hermitian code will correct at least $t = \frac{n - \alpha - 1}{2}$ errors. In this section, we will use the Hermitian curve with the parameter $q = 16$, $r = 2$, and $u = 17$ in Table 1. That is, we will work on the finite field $\mathbb{F}_{2^8}$ and the Hermitian curve contains $2^{12} = 4096$ elements with $g = 120$. Table 2 lists the parameters for HermitianRLCE encryption scheme at the security levels 128, 192, and 256 bits where κ_c is the classical security level and κ_q is the quantum security level. As a comparison, we also include the corresponding public key size for Generalized Reed-Solomon code based RLCE schemes. It is noted that for security level 128 and 192, HermitianRLCE's public key size is approximiately 80% of the GRS-RLCE public key size. For the security level 256, HermitianRLCE's public key size is approximiately 72% of the GRS-RLCE public key size.

7   HermitianRLCE with Other Extended Norm-Trace Curves

In Sect. 6, we proposed HermitianRLCE parameters for Hermitian curves with $q = 16$, $r = 2$, and $u = 17$. It would be interesting to know whether other curves have advantages in reducing the public key sizes and improve the encryption/decryption performance. A product Hermitian code has dimension $2k + g - 1$ which is "closer" to the product code dimension of a random code (compared with the dimension $2k - 1$ for the product code of a GRS code). Thus smaller values w in HermitianRLCE schemes are sufficient to defeat filtration attacks. The smaller choice of w has significantly reduced the public key size of HermitianRLCE schemes (compared with GRS-RLCE schemes). However, the value of w should be sufficiently larger so that $\binom{n+w}{w} \geq 2^{128}$ (respectively 2^{192} and 2^{256}) for the security level of AES-128. For the parameter $q = 4$, $r = 4$, and $u = 5$, Table 3 lists the parameters for HermitianRLCE encryption scheme at the security levels

Table 3. Parameters for HermitianRLCE scheme with $q = 4$, $r = 4$, and $u = 5$

ID	κ_c	κ_q	n	k	t	w	cipher bytes	sk	pk bytes	GRS-RLCE pk bytes
0	128	84	640	295	110	22	640		108265	188001
1	192	118	870	435	155	33	870		203580	450761

Table 4. Parameters for HermitianRLCE scheme with $q = 4$, $r = 4$, and $u = 17$

ID	κ_c	κ_q	n	k	t	w	cipher bytes	sk	pk bytes	GRS-RLCE pk bytes
0	128	84	1270	357	205	22	1270		333795	188001
1	192	118	1540	537	250	33	1540		556332	450761
2	256	148	1810	687	310	45	1810		802416	1232001

128 and 192. For this parameter set, it is not possible to choose a parameter sets for the 256-bit security level since the Hermitian curve contains 1024 points. The public key size is relatively larger for the parameters $q = 4$, $r = 4$, and $u = 5$.

As another example, we analyze security levels for HermitianRLCE schemes based on extended norm-trace curves with parameter $q = 4$, $r = 4$, and $u = 17$. Table 4 lists the parameters for HermitianRLCE encryption schemes at the security levels 128, 192, and 256. The public key size is significantly larger for the parameters $q = 4$, $r = 4$, and $u = 17$.

Our analysis in Tables 2, 3 and 4 shows that other extended norm-trace curves with smaller genus can be used to build HermitianRLCE schemes with smaller public key sizes. Thus the preferred extended norm-trace for HermitianRLCE encryption schemes are based on Hermitian curves with $q = 16$, $r = 2$, and $u = 17$

References

1. Baldi, M., Bodrato, M., Chiaraluce, F.: A new analysis of the McEliece cryptosystem based on QC-LDPC codes. In: Ostrovsky, R., De Prisco, R., Visconti, I. (eds.) SCN 2008. LNCS, vol. 5229, pp. 246–262. Springer, Heidelberg (2008). https://doi.org/10.1007/978-3-540-85855-3_17
2. Couvreur, A., Lequesne, M., Tillich, J.-P.: Recovering short secret keys of RLCE in polynomial time. arXiv preprint arXiv:1805.11489 (2018)
3. Couvreur, A., Otmani, A., Tillich, J.P.: Polynomial time attack on wild McEliece over quadratic extensions. In: Nguyen, P.Q., Oswald, E. (eds.) EUROCRYPT 2014. LNCS, vol. 8441, pp. 17–39. Springer, Heidelberg (2014). https://doi.org/10.1007/978-3-642-55220-5_2
4. Couvreur, A., Márquez-Corbella, I., Pellikaan, R.: A polynomial time attack against algebraic geometry code based public key cryptosystems. In: 2014 IEEE International Symposium on Information Theory (ISIT), pp. 1446–1450. IEEE (2014)
5. Grassl, M., Langenberg, B., Roetteler, M., Steinwandt, R.: Applying Grover's algorithm to AES: quantum resource estimates. In: Takagi, T. (ed.) PQCrypto 2016. LNCS, vol. 9606, pp. 29–43. Springer, Cham (2016). https://doi.org/10.1007/978-3-319-29360-8_3
6. Janwa, H., Moreno, O.: Mceliece public key cryptosystems using algebraic-geometric codes. Des., Codes Cryptogr. **8**(3), 293–307 (1996)
7. Kepley, S., Steinwandt, R.: Quantum circuits for F_{2^m}-multiplication with subquadratic gate count. Quantum Inf. Process. **14**(7), 2373–2386 (2015)

8. McEliece, R.J.: A public-key cryptosystem based on algebraic coding theory. DSN Prog. Rep. **42**(44), 114–116 (1978)
9. Misoczki, R., Tillich, J.-P., Sendrier, N., Barreto, P.: MDPC-McEliece: new McEliece variants from moderate density parity-check codes. In: Proceedings of IEEE ISIT 2013, pp. 2069–2073 (2013)
10. Peters, C.: Information-set decoding for linear codes over $\mathbf{F}_q$. In: Sendrier, N. (ed.) PQCrypto 2010. LNCS, vol. 6061, pp. 81–94. Springer, Heidelberg (2010). https://doi.org/10.1007/978-3-642-12929-2_7
11. Stern, J.: A method for finding codewords of small weight. In: Cohen, G., Wolfmann, J. (eds.) Coding Theory 1988. LNCS, vol. 388, pp. 106–113. Springer, Heidelberg (1989). https://doi.org/10.1007/BFb0019850
12. Wang, Y.: Quantum resistant random linear code based public key encryption scheme RLCE. In: Proceedings of IEEE ISIT, pp. 2519–2523, July 2016
13. Wang, Y.: Revised quantum resistant public key encryption scheme RLCE and IND-CCA2 security for McEliece schemes. IACR ePrint, July 2017. https://eprint.iacr.org/2017/206.pdf
14. Yang, K., Kumar, P.V.: On the true minimum distance of Hermitian codes. In: Stichtenoth, H., Tsfasman, M.A. (eds.) Coding Theory and Algebraic Geometry, pp. 99–107. Springer, Heidelberg (1992). https://doi.org/10.1007/BFb0087995

LEDAcrypt: QC-LDPC Code-Based Cryptosystems with Bounded Decryption Failure Rate

Marco Baldi[1]([✉]), Alessandro Barenghi[2], Franco Chiaraluce[1], Gerardo Pelosi[2], and Paolo Santini[1]

[1] Università Politecnica delle Marche, Ancona, Italy
{m.baldi,f.chiaraluce}@univpm.it, p.santini@pm.univpm.it
[2] Politecnico di Milano, Milano, Italy
{alessandro.barenghi,gerardo.pelosi}@polimi.it

Abstract. We consider the QC-LDPC code-based cryptosystems named LEDAcrypt, which are under consideration by NIST for the second round of the post-quantum cryptography standardization initiative. LEDAcrypt is the result of the merger of the key encapsulation mechanism LEDAkem and the public-key cryptosystem LEDApkc, which were submitted to the first round of the same competition. We provide a detailed quantification of the quantum and classical computational efforts needed to foil the cryptographic guarantees of these systems. To this end, we take into account the best known attacks that can be mounted against them employing both classical and quantum computers, and compare their computational complexities with the ones required to break AES, coherently with the NIST requirements. Assuming the original LEDAkem and LEDApkc parameters as a reference, we introduce an algorithmic optimization procedure to design new sets of parameters for LEDAcrypt. These novel sets match the security levels in the NIST call and make the C99 reference implementation of the systems exhibit significantly improved figures of merit, in terms of both running times and key sizes. As a further contribution, we develop a theoretical characterization of the decryption failure rate (DFR) of LEDAcrypt cryptosystems, which allows new instances of the systems with guaranteed low DFR to be designed. Such a characterization is crucial to withstand recent attacks exploiting the reactions of the legitimate recipient upon decrypting multiple ciphertexts with the same private key, and consequentially it is able to ensure a lifecycle of the corresponding key pairs which can be sufficient for the wide majority of practical purposes.

1 Introduction

In this work, we provide theoretical and implementation advancements concerning quasi-cyclic low-density parity-check (QC-LDPC) code-based cryptosystems known as LEDAcrypt [3], which are under consideration by NIST for the

The work of Paolo Santini was partially supported by Namirial S.p.A.

M. Baldi et al. (Eds.): CBC 2019, LNCS 11666, pp. 11–43, 2019.
https://doi.org/10.1007/978-3-030-25922-8_2

second round of standardization of post-quantum cryptographic systems [31]. These new systems are built upon two previous systems named LEDAkem (low-density parity-check code-based key encapsulation mechanism) and LEDApkc (low-density parity-check code-based public-key cryptosystem), which were submitted to the first round of the NIST competition.

The mathematical problem on which these systems rely is the one of decoding a random-looking linear block code. Such a problem belongs to the class of NP-complete problems [6,22], which is known to contain problems without polynomial time solution on a quantum computer. This line of research was initiated by McEliece in 1978 [27], using Goppa codes as secret codes, and Niederreiter in 1986 [32], with a first attempt of introducing generalized Reed-Solomon (GRS) codes in such framework. With the main aim of reducing the public key size, several other families of codes have been considered during years, like quasi-cyclic (QC) codes [12], low-density parity-check (LDPC) codes [30], quasi-dyadic (QD) codes [28], QC-LDPC codes [4] and quasi-cyclic moderate-density parity-check (QC-MDPC) codes [29].

The distinguishing points of the LEDAcrypt cryptosystems with respect to other code-based post-quantum cryptosystems relies on the use of QC-LDPC codes as secret codes and on an efficient decoding algorithm recently introduced for codes of this kind [2]. The two main attacks that can be mounted against these systems are a decoding attack (DA) and a key recovery attack (KRA) both exploiting information set decoding (ISD) algorithms. In addition, recent attacks based on the information leakage arising from the observation of the reaction of someone decrypting ciphertexts with the same private key have proved to be effective in reducing the lifecycle of keypairs used in LEDApkc and other code-based cryptosystems characterized by a non-zero DFR [9,10,16].

In this work, we analyze all the aforementioned attacks and show how to tune the parameter design of LEDAcrypt in order to foil them. Our contributions can be summarized as follows.

(i) A quantification of the quantum and classical computational efforts required to break the Advanced Encryption Standard (AES) is provided. We rely on typical circuit design estimates for the classical computing complexity, and on the work by Grassl et al. [14] for the quantum computing complexity.

(ii) A new algorithmic approach to the design of LEDAcrypt instances with parameters matching the NIST requirements, is introduced. The proposed approach employs finite regime estimations (as opposed to asymptotic bounds [21]) of the computational efforts required to perform ISD attacks as well as to execute an exhaustive search in the parameter space of the algorithms. The parameters designed through this method yield key sizes which are significantly smaller than the original proposal to the NIST standardization effort.

(iii) A novel, closed-form upper bound on the DFR of the LEDAcrypt instances is provided. This allows to include the DFR as constraint of the parameter design, to generate keypairs with a sufficiently small DFR to provide security against chosen ciphertext attacks, such as reaction attacks. We also report sample sets of parameters targeting a DFR of 2^{-64} for long term keys in LEDAcrypt, as well as a DFR smaller than $2^{-\lambda}$, where λ equals 128, 192 and 256, for the NIST security categories 1, 3 and 5, respectively.

The paper is organized as follows. In Sect. 2 we briefly recall the LEDAcrypt systems and the relevant notation. In Sect. 3 we define the security level benchmarks we consider, in compliance with the NIST requirements. In Sect. 4 we describe the attacks we take into account in the system design. In Sect. 5 we describe an algorithmic procedure for the design of optimized sets of parameters for these systems. In Sect. 6 we present the parameter sets resulting from the algorithmic design procedure for keypairs to be used in ephemeral key encapsulation mechanism (KEM)s. In Sect. 7 we report parameter sets that guarantee a DFR lower than a given threshold for LEDAcrypt instances with long term keys and indistinguishability under adaptive chosen ciphertext attack (IND-CCA2). The latter are derived on the basis of a theoretical characterization of the DFR of LEDAcrypt, which is reported in Appendix A. Finally, in Sect. 8 we draw some conclusive remarks.

2   Preliminaries and Notation

LEDAcrypt exploits a secret key (SK) formed by two binary matrices: H is the binary parity-check matrix of a secret QC-LDPC code and Q is a secret transformation matrix. The code described by H has length $n = pn_0$ and dimension $k = p(n_0 - 1)$, where p is a large prime integer and n_0 is a small integer. The matrix H is formed by a row of n_0 circulant matrices with size $p \times p$ and weight d_v. The matrix Q is formed by $n_0 \times n_0$ circulant matrices whose weights coincide with the entries of $\bar{m} = [m_0, m_1, \ldots, m_{n_0-1}]$ for the first row and with those of cyclically shifted versions of $\bar{m}$ for the subsequent rows. Both H and Q are sparse, and their product gives a sparse matrix $H' = HQ$ that is a valid parity-check matrix of the public code. Due to its sparsity, H' cannot be disclosed, thus the public key is a linearly transformed version of H' that hides its sparsity. The LEDAcrypt cryptosystems hide the LDPC structure of H' multiplying all its circulant blocks by the multiplicative inverse of the last block of H' itself, yielding the public key matrix pk^{Nie}.

Concerning the error correction capability of these codes, we recall that classical hard-decision decoding algorithms used for QC-LDPC codes are known as Bit Flipping (BF) decoders. The LEDA cryptosystems employ a different decoding strategy which, while retaining a fixed point BF approach, is more efficient than the schoolbook BF. Such a procedure, known as Q-decoder [2], relies on the fact that the (secret) parity-check matrix $sk^{\text{Nie}} = H'$ is obtained as the product of two sparse matrices, i.e., $H' = HQ$, where H has size $(n-k) \times n$ and number of non-zero elements in a row equal to $d_c = n_0 d_v \ll n$, while Q has size $n \times n$ and number of non-zero elements in a column equal to $m = \sum_i m_i \ll n$. Both the BF and the Q-decoder are not bounded-distance decoders, therefore their decoding radius cannot be easily ascertained for a given, weight-t, error, resulting in a non-null decoding failure rate in practical scenarios. We denote as $t \ll n$ the number of errors that can be corrected by the code defined by H' with a sufficiently high probability, and the code itself is denoted as $C(n, k, t)$.

Given a value for t, the encryption is performed encoding the secret message with the public code followed by the addition of a weight t error vector (in the McEliece setting) or mapping the secret message into a weight-t binary error vector and computing its syndrome through the public code (in the Niederreiter setting). The decryption of LEDAcrypt KEM performs syndrome decoding on the received syndrome after multiplying it by the last circulant block of $sk^{\mathtt{Nie}} = H'$. The decoding retrieves the error vector e except for the cases where a decoding failure takes place.

LEDAcrypt provides a KEM, named LEDAcrypt KEM, employing the Niederreiter cryptosystem with One Wayness against Chosen Plaintext Attack (OW-CPA) and an apt conversion to obtain a KEM with a twofold goal: (i) provide a fast KEM with indistinguishability under chosen plaintext attack (IND-CPA) and ephemeral keys for low latency session establishment with perfect forward secrecy, and (ii) provide a KEM with IND-CCA2 and long term keys for scenarios where key reuse may be desirable. We achieve this goal employing the same IND-CCA2 conversion applied to the QC-LDPC Niederreiter cryptosystem for both scenarios and achieving an appropriate DFR through code parameter tuning and, where needed, an additional IND-CCA2 redundant encryption technique. In particular, we employ the $U_m^{\not\perp}$ construction defined in [18], which starts from a deterministic cryptosystem to build a KEM with IND-CCA2 in the Random Oracle Model (ROM) with a tight reduction.

The same construction was proven to achieve IND-CCA2 in the Quantum Random Oracle Model (QROM) in [19], with a tighter security reduction being reported in [20], starting from the assumption that the underlying deterministic cryptosystem is OW-CPA, as it is the case with our Niederreiter KEM. The proofs in [18–20] take into account the possibility that the underlying cryptoscheme is characterized by a bounded correctness error δ. The instantiation of the $U_m^{\not\perp}$ construction employing the QC-LDPC code-based Niederreiter cryptoscheme and a cryptographically secure hash, $\mathtt{Hash}(\cdot)$, is reported in Fig. 1. We chose, as the cryptographically secure hash to instantiate the $U_m^{\not\perp}$ construction, the NIST standard SHA-3 hash function.

In case of a decoding failure [18–20], the decapsulation procedure computes the returned outcome by hashing a secret value and the ciphertext. This prevents an adversary from distinguishing when a failure occurs due to malformed plaintext messages, i.e., messages with a number of asserted bits that is not exactly equal to t, from when a failure occurs due to the intrinsic behavior of the underlying QC-LDPC code. In other terms, the adversary cannot draw any conclusion about the decoding abilities of the code at hand when he/she is in control of composing messages that are not in the legitimate message space.

To provide IND-CCA2 for a given security level 2^λ, the authors of [18] state that it is required for the decryption function to have a correctness error $\delta \leq 2^{-\lambda}$.

Algorithm 1: LEDAcrypt-KEM ENCAP

Input: pk^{Nie}: public key.

Output: c: encapsulated ephemeral key;
    K: ephemeral key.

Data: $p>2$ prime, $\text{ord}_p(2)=p-1$, $n_0 \geq 2$;
    ENCRYPT$^{\text{Nie}}(e, pk^{\text{Nie}})$: encryption of the Niederreiter cryptosystem;
    $\mathcal{E}$: set of all possible binary error vectors $e=[e_0|\ldots|e_{n_0-1}]$, $wt(e)=t$

1   $e \xleftarrow{\$} \mathcal{E}$ `// uniform random picking`
2   $c \leftarrow$ ENCRYPT$^{\text{Nie}}(e, pk^{\text{Nie}})$
3   $K \leftarrow \text{Hash}(e)$

4 **return** (c, K)

(a)

Algorithm 2: LEDAcrypt-KEM DECAPS

Input: sk^{Nie}: secret key
    k: a secret random bitstring;
    c: encapsulated key.

Output: K: decapsulated key.

Data: $p>2$ prime, $\text{ord}_p(2)=p-1$, $n_0 \geq 2$;
    DECRYPT$^{\text{Nie}}(c, sk^{\text{Nie}})$: decryption function returning **res** = **false** on an
    incorrect decoding, **true** and the original message e, otherwise.

1   $\{e, \textbf{res}\} \leftarrow$ DECRYPT$^{\text{Nie}}(c, sk^{\text{Nie}})$
2   **if res** = **true** and $wt(e) = t$ **then**
3   **return** Hash(e)
4   **else**
5   **return** Hash$(c|k)$

(b)

Fig. 1. Key encapsulation (a) and key decapsulation (b) primitives of LEDAcrypt KEM

Given our goal of having both a fast and compact KEM with ephemeral keys and IND-CPA guarantees, as well as a KEM with IND-CCA2, we will provide different sets of parameters to be employed in the LEDAcrypt KEM construction, which are characterized by a DFR low enough to foil statistical attacks and achieve IND-CCA2 guarantees, without hindering practical deployment.

In addition to the LEDAcrypt KEM, LEDAcrypt provides a public-key cryptosystem (PKC) with IND-CCA2 guarantees. While it is possible to employ LEDAcrypt KEM in a Key Encapsulation Module + Data Encapsulation Mechanism (KEM+DEM) combination with a symmetric encryption primitive, we note that such an approach may lead to a non-negligible ciphertext expansion in case plaintexts are small in size. To overcome such an issue, LEDAcrypt PKC provides a construction that starts from the QC-LDPC code-based McEliece cryptosystem and derives a PKC exploiting the available capacity of the McEliece Public-Key Encryption (PKE) primitive to store the actual message content. It is worth noting that, in the McEliece setting, the systematic form of the generator matrix of the public code included in the public key would easily allow any observer to recover the information word embedded in an encrypted message, without recovering the private key. Nevertheless, the conversion proposed by Kobara and Imai in [23], with the purpose of maximizing the amount of messages encrypted by a McEliece PKC, allows IND-CCA2 guarantees to be provided in the ROM. Therefore, the confidentiality of the information word as well as the security of the private key remain guaranteed by the hardness of the NP-hard general decoding problem even when a systematic generator matrix is employed as public key. For a detailed description of the basic encryption and decryption transformations of LEDAcrypt PKC, as well as the mechanisms of the γ-conversion scheme [23] that allow us to obtain an IND-CCA2 version of LEDAcrypt PKC, we refer the reader to the LEDAcrypt specification [3].

3   Security Level Goals

The bar to be cleared to design parameters for post-quantum cryptosystems was set by NIST to the computational effort required on either a classical or a quantum computer to break the AES with a key size of λ bits, $\lambda \in \{128, 192, 256\}$, through an exhaustive key search. The three pairs of computational efforts required on a classical and quantum computer correspond to NIST Category 1, 3, and 5, respectively [31]. Throughout the design of the parameters for LEDAcrypt we ignore Categories 2 and 4: if a cipher matching those security levels is required, we advise to employ the parameters for Categories 3 and 5, respectively.

The computational worst-case complexity of breaking AES on a classical computer can be estimated as $2^\lambda C_{AES}$, where C_{AES} is the amount of binary operations required to compute AES on a classical computer on a small set of plaintexts, and match them with a small set of corresponding ciphertexts to validate the correct key retrieval. Indeed, more than a single plaintext-ciphertext pair is required to retrieve AES keys [14]. In particular, a validation on three plaintext-ciphertext pairs should be performed for AES-128, on four pairs for AES-192 and on five for AES-256.

Willing to consider a realistic AES implementation for exhaustive key search purposes, we refer to [38], where the authors survey the state of the art of Application-Specific Integrated Circuit (ASIC) AES implementations, employing the throughput per Gate Equivalent (GE) as their figure of merit. The most performing AES implementations are the ones proposed in [38], and require around 16ki GEs. We thus deem reasonable to estimate the computational complexity of an execution of AES as 16ki binary operations. We are aware of the fact that this is still a conservative estimate, as we ignore the cost of the interconnections required to carry the required data to the AES cores.

The computational complexity of performing an AES key retrieval employing a quantum computer was measured first in [14], where a detailed implementation

Table 1. Classical and quantum computational costs of a key search on AES

NIST category	AES key size (bits)	Classical cost (binary operations)	Quantum cost [14] (quantum gates)
1	128	$2^{128} \cdot 2^{14} \cdot 3 = 2^{143.5}$	$1.16 \cdot 2^{81}$
3	192	$2^{192} \cdot 2^{14} \cdot 4 = 2^{208}$	$1.33 \cdot 2^{113}$
5	256	$2^{256} \cdot 2^{14} \cdot 5 = 2^{272.3}$	$1.57 \cdot 2^{145}$

of an AES breaker is provided. The computation considers an implementation of Grover's algorithm [15] seeking the zeros of the function given by the binary comparison of a set of AES ciphertexts with the encryption of their corresponding plaintexts for all the possible key values. The authors of [14] chose to report the complexity of the quantum circuit computing AES counting only the number of the strictly needed Clifford and T gates, since they are the ones currently most expensive to implement in practice. Selecting a different choice for the set of quantum gates employed to realize the AES circuit may yield a different complexity; however, the difference will amount to a reasonably small constant factor, as it is possible to re-implement the Clifford and T gates at a constant cost with any computationally complete set of quantum gates. We thus consider the figures reported in [14] as a reference for our parameter design procedure. In Table 1 we summarize the computational cost of performing exhaustive key searches on all three AES variants (i.e., with 128, 192, and 256 bits long keys), both considering classical and quantum computers.

4 Evaluated Attacks

Let us briefly recall the set of attacks to be considered in the design of the system parameters. In addition to advanced attacks, we also consider some basic attack procedures, such as exhaustive key search, which must be taken into account in any automated cryptosystem parameter optimization, since they impose some bounds on the system parameters.

An open source software implementation of the routines for computing the complexity of the described attacks is publicly available [1].

4.1 Attacks Based on Exhaustive Key Search

Enumerating all the possible values for the secret key is, in principle, applicable to any cryptosystem. The original LEDAkem and LEDApkc specification documents do not mention exhaustive key search, as it is possible to verify that they are strictly dominated by other, less computationally demanding, attack strategies such as the use of ISD algorithms.

In this parameter revision, in order to pose suitable bounds to the automated parameter search we perform, we consider the application of an exhaustive enumeration strategy to each one of the two secret low-density binary matrices constituting the LEDAcrypt secret keys, i.e., H and Q. We recall that H is a block circulant binary matrix constituted by $1 \times n_0$ circulant blocks with size equal to p bits, where $n_0 \in \{2, 3, 4\}$ and p is a prime such that $\mathrm{ord}_2(p) = p - 1$ (i.e., $2^{p-1} \bmod p = 1 \bmod p$). Q is a binary block circulant matrix constituted by $n_0 \times n_0$ binary circulant blocks with size p. Willing to follow a conservative approach, we design revised parameter sets such that it is not possible for an attacker to enumerate all the possible matrices H or Q. While there is no standing attack benefiting from such an enumeration, we deem reasonable adding such a constraint to the design of the parameter sets as a peace-of-mind measure. In our approach, to prevent attacks relying on the exhaustive search for the value of either H or Q, we considered the remainder of the attack strategy which may be employed to derive the matrix not being exhaustively searched for to have a constant complexity (i.e. $\Theta(1)$). This in turn implies that any attack strategy which leverages the exhaustive search of H or Q to obtain useful information for a key recovery attack will have a computational complexity matching or exceeding the required security level.

Considering that each row of a circulant block of H has Hamming weight d_v, a straightforward counting argument yields $\sharp H = \binom{p}{d_v}^{n_0}$ as the number of possible choices for H. The number of possible choices for Q, denoted as $\sharp Q$, can be derived starting from the consideration that the weights of a row of each circulant block in a block-row of Q are equal for all the rows up to a circular shift. Such weights, denoted as $\{m_0, \ldots, m_{n_0-1}\}$, allow to write the number of possible choices for Q as $\sharp Q = \left[\prod_{i \in \{m_0, \ldots, m_{n_0-1}\}} \binom{p}{i} \right]^{n_0}$.

Considering the case where the key recovery strategy exploits the enumeration of either H or Q within an algorithm running on a quantum computer, we consider the possibility of employing a Grover-like strategy to speedup the enumeration of either H or Q. Assuming conservatively that such a strategy exists, we consider the resistance against exhaustive key search with a quantum computer to be $\sqrt{\sharp H}$ and $\sqrt{\sharp Q}$ for the search over H and Q, respectively. We note that, for all parameter sets proposed in the original specification [3], the cost of enumerating H and Q exceeds that of the best attacks via ISD.

4.2   Attacks Based on Information Set Decoding

It is well known that efficient message and key recovery attacks against McEliece and Niederreiter cryptosystem variants based on LDPC and MDPC codes are those exploiting ISD algorithms. Such algorithms have a long development history, dating back to the early '60s [34], and provide a way to recover the error pattern affecting a codeword of a generic random linear block code given a representation of the code in the form of either its generator or parity-check matrix.

Despite the fact that the improvement provided by ISD over the straightforward enumeration of all the possible error vectors affecting the codeword is only polynomial, employing ISD provides substantial speedups. It is customary for ISD variant proposers to evaluate the effectiveness of their attacks considering the improvement on a worst-case scenario as far as the code rate and number of corrected errors goes (see, for instance [5]). Such an approach allows deriving the computational complexity as a function of a single variable, typically taken to be the code length n, and obtaining asymptotic bounds for the behavior of the algorithms. In our parameter design, however, we chose to employ non-asymptotic estimates of the computational complexity of the ISD attacks. Therefore, we explicitly compute the amount of time employing a non-asymptotic analysis of the complexity of ISD algorithms, given the candidate parameters of the code at hand. This approach also permits us to retain the freedom to pick rates for our codes which are different from the worst-case one for decoding, thus exploring different trade-offs in the choice of the system parameters. In case the ISD algorithm has free parameters, we seek the optimal case by explicitly computing the complexity for a large region of the parameter space, where the minimum complexity resides. We consider the ISD variants proposed by Prange [34], Lee and Brickell [24], Leon [25], Stern [36], Finiasz and Sendrier [11], and Becker, Joux, May and Meurer (BJMM) [5], in our computational complexity evaluation on classical computers. The reason for considering all of them is to avoid concerns on whether their computational complexity in the finite-length regime is already well approximated by their asymptotic behavior. In order to estimate the computational complexity of ISD on quantum computing machines, we consider the results reported in [8], which are the same employed in the original specification [3]. Since complete and detailed formulas are available only for the ISD algorithms proposed by Lee and Brickell, and Stern [36], we consider those as our computational complexity bound. While asymptotic bounds show that executing a quantum ISD derived from the May-Meurer-Thomae (MMT) algorithm [26] is faster than a quantum version of Stern's [21], we note that there is no computational complexity formulas available for generic code and error rates.

Message Recovery Attacks Through ISD. ISD algorithms can effectively be applied to recover the plaintext message of any McEliece or Niederreiter cryptosystem instance by retrieving the intentional error pattern used during encryption. When a message recovery attack of this kind is performed against a system variant exploiting quasi cyclic codes, like those at hand, it is known that a speedup equal to the square root of the circulant block size can be achieved [35]. We consider such message recovery attacks in our parameter design, taking this speedup into account in our computations.

Key Recovery Attacks Through ISD. The most efficient way, and currently the only known way, to exploit the sparsity of the parity checks that characterizes the secret code $H' = HQ$ in order to attack LEDAcrypt is trying to recover a low-weight codeword of the dual of the public code. In fact, any sparse row of H' is a low-weight codeword belonging to the dual of the public code, and such codewords have a weight that is very close or equal to $d' = n_0 d_v(\sum_{i=0}^{n_0-1} m_i)$, which is comparatively small with respect to the codeword length n.

Therefore, it is possible to search for such low-weight codewords through ISD algorithms, which are far more efficient than trying all the $\binom{n}{d'}$ possible codewords. Indeed, the complexity of accomplishing this task through ISD is equal to the one of decoding a code of the same length n, with dimension equal to the redundancy $r = n - k$ of the code at hand, and with d' errors.

We consider such key recovery attacks in our parameter design, evaluating their complexity for all the aforementioned ISD algorithms.

4.3   Reaction Attacks

In addition to the proper sizing of the parameters of LEDAcrypt so that it withstands the aforementioned attacks, a last concern should be taken into account regarding the lifetime of a LEDAcrypt key pair, when keys are not ephemeral. In fact, whenever an attacker may gain access to a decryption oracle to which he may pose a large amount of queries, the so-called *reaction attack* becomes applicable. Reaction attacks recover the secret key by exploiting the inherent non-zero DFR of QC-LDPC codes [9,10,17]. In particular, these attacks exploit the correlation between the DFR of the code, the positions of the parity checks in the private matrix, and the error positions in the error vector. Indeed, whenever e and either H' have pairs of ones placed at the same distances, the decoder exhibits a DFR smaller than the average.

Such attacks require the collection of the outcome of decoding (success or failure) on a ciphertext for which the attacker knows the distances in the support of the error vector, for a significant number of ciphertexts, to achieve statistical confidence in the result. The information on the decoding status is commonly referred to as the *reaction* of the decoder, hence the name of the attack. The strongest countermeasure against these attacks is to choose a proper set of system parameters such that the DFR is negligible, which means an attacker would require a computationally unfeasible amount of decryption actions to obtain even a single decoding failure. It has been recently pointed out in [33] that some mechanisms exist to generate error vectors able to artificially increase the DFR of systems such as LEDAcrypt. However, such techniques require to known an error vector causing a single decoding failure to be carried out. Therefore, choosing an appropriately low DFR such attacks can be made as expensive as a key recovery via ISD. In addition, these methods require the manipulation of error vectors, which is not feasible when an IND-CCA2 secure conversion is adopted.

Instances with Ephemeral Keys and Accidental Key Reuse. LEDAcrypt KEM instances with Perfect Forward Secrecy (PFS) and IND-CPA exploit ephemeral keys that are renewed before each encryption. Hence, any key pair can be used to decrypt one ciphertext only. In such a case, statistical attacks based on the receivers' reactions are inherently unfeasible on condition that the ephemeral nature of the keys is strictly preserved.

Reaction attacks could instead be attempted in the case of an accidental reuse of the keys in these instances. However, the parameter choices of LEDAcrypt KEM instances with ephemeral keys guarantee a DFR in the order of 10^{-8}–10^{-9}. An attacker would need to collect DFR^{-1} ciphertexts encrypted with the same key, on average, before observing one decryption failure. Hence, these instances are protected against a significant amount of accidental key reuse. Moreover such a low DFR provides a good practical reliability for the ephemeral KEM, as it is very seldom needed to repeat a key agreement due to a decoding failure.

Instances with Long Term Keys. LEDAcrypt KEM and LEDAcrypt PKC instances with long term keys employ a suitable conversion to achieve IND-CCA2. The IND-CCA2 model assumes that an attacker is able to create a polynomially bound number of chosen ciphertexts and ask for their decryption to an oracle owning the private key corresponding to the public key used for their generation. Note that such an attacker model includes reaction attacks, since the adversary is able to observe a large number of decryptions related to the same keypair.

A noteworthy point is that the current existing IND-CCA2 constructions require the DFR of the scheme to be negligible. Indeed, most IND-CCA2 attaining constructions require the underlying cryptosystem to be correct, i.e., $D_{sk}(E_{pk}(m)) = m$, for all valid key pairs (pk, sk) and for all valid messages m. Recent works [19, 20] tackled the issue of proving a construction IND-CCA2 even in the case of an underlying cipher affected by decryption failures. The results obtained show that, in case the DFR is negligible in the security parameter, it is possible for the construction to attain IND-CCA2 guarantees even in case of decryption failures.

In systems with non-zero DFR, endowed with IND-CCA2 guarantees, attacks such as crafting a ciphertext aimed at inducing decoding errors are warded off. Therefore, our choice of employing an IND-CCA2 achieving construction to build both our PKC and KEM, paired with appropriate parameters guaranteeing a negligible DFR allows us to thwart ciphertext alteration attacks such as the ones pointed out in the official comments to the first round of the NIST competition[1].

5 Parameter Design

In this section we describe an automated procedure for the design of parameters for the QC-LDPC codes employed in LEDAcrypt. An open source software implementation of the routines for computing the complexity of the described attacks and perform parameter generation is available as public domain software at [1]. The LEDAcrypt design procedure described in this section takes as input the desired security level λ_c and λ_q, expressed as the base-2 logarithm of the number of operations of the desired computational effort on a classical and quantum computer, respectively. In addition to λ_c and λ_q, the procedure also takes as input the number of circulant blocks, $n_0 \in \{2, 3, 4\}$, forming the parity-check matrix H, allowing tuning of the code rate. As a third and last parameter, the procedure takes as input the value of ϵ, which tunes the estimate of the

---

[1] https://csrc.nist.gov/CSRC/media/Projects/Post-Quantum-Cryptography/documents/round-1/official-comments/LEDAkem-official-comment.pdf.

system DFR for the IND-CPA case. We first consider instances with ephemeral keys, which are designed using $\epsilon = 0.3$: the resulting DFR values are in the range 10^{-9}–10^{-8}. The design of parameters for long term keys starts from the output of the procedure employed for ephemeral key parameters, increasing the value of p until the bounds specified in Sect. 7 provide a sufficiently small DFR.

The parameter design procedure outputs the size of the circulant blocks, p, the weight of a column of H, d_v, the number of intentional errors, t, the weights of the n_0 blocks of a row of Q, i.e., $\{m_0, m_1, \ldots, m_{n_0-1}\}$, with $\sum_{i=0}^{n_0-1} m_i = m$. The procedure enforces the following constraints on the parameter choice:

- Classical and quantum exhaustive searches for the values of H or Q should require at least 2^{λ_c} and 2^{λ_q} operations. This constraint binds the value of the circulant block size p and the weight of a row of the circulant block, d_v for H and m_i for Q, to be large enough.
- The minimum cost for a message recovery via ISD on both quantum and classical computers must exceed 2^{λ_q} and 2^{λ_c} operations, respectively. This constraint binds the values of the code length $n = n_0 p$, the code dimension $k = (n_0 - 1)p$ and the number of errors t to be chosen such that an ISD on the code $\mathcal{C}(n, k, t)$ requires more than 2^{λ_q} or 2^{λ_c} operations on a quantum and a classical computer.
- The minimum cost for a key recovery attack via ISD on both quantum and classical computers must exceed 2^{λ_q} and 2^{λ_c} operations, respectively. This constraint binds the values of the code length $n = n_0 p$, the code redundancy $r = p$ and the number of ones in a row of HQ, $d'_v n_0$, with $d'_v = d_v m$ to be chosen such that an ISD on the code $\mathcal{C}(n, r, d'_v n_0)$ requires more than 2^{λ_q} or 2^{λ_c} operations on a quantum and classical computer, respectively.
- The choice of the circulant block size, p, should be such that p is a prime number and $\mathrm{ord}_2(p) = p - 1$ in order to ensure non-singularity of the last block of H'.
- The choice of the circulant block size, p, and parity-check matrix density, $n_0 d_v$, must allow the code to correct the required amount of errors. This is tested through the computation of the decoding threshold, as described in the original specification [3].
- The weights of the circulant blocks of Q must be such that the permanent of the matrix of the block weights is odd, which guarantees the existence of its multiplicative inverse (see the full LEDAcrypt specification [3] for details).

We report a synthetic description of the procedure implemented in the publicly available code as Algorithm 3. The rationale of the procedure[2] is to proceed in refining the choice for p, t, d_v, and all the m_i's at fixed point, considering only values of p respecting $\mathrm{ord}_2(p) = p - 1$.

---

[2] Note that, in the pseudocode of Algorithm 3, the loop construct **while**($<$ condition $>$) ... iterates the execution of instructions in the loop body when the condition is **true**, while the loop construct **Repeat** ... **until**($<$ condition $>$) iterates the instructions in the loop body when the condition is **false**.

Algorithm 3: LEDAcrypt Parameter Generation

Input: λ_c, λ_q:desired security levels against classical and quantum attacks, respectively;
ϵ: safety margin on the minimum size of the secret parity-check matrix H, named $p_{th} = p(1 + \epsilon)$, where p is the size of a circulant block, so that the code is expected to correct all the errors with acceptable DFR;
n_0: number of circulant blocks of the $p \times n_0 p$ parity-check matrix H of the code. The Q matrix is constituted by $n_0 \times n_0$ circulant blocks as well, each of size p.

Output: p: size of a circulant block; t: number of errors; d_v: weight of a column of the parity matrix H; $\langle m_0, m_1, \ldots, m_{n_0-1} \rangle$: an integer partition of m, the weight of a row of the matrix Q. Each m_i is the weight of a block of Q.

Data: NEXTPRIME(x): subroutine returning the first prime p larger than the value of the input parameter and such that $\text{ord}_2(p) = p - 1$;
C-ISD-COST(n, k, t), Q-ISD-COST(n, k, t): subroutines returning the costs of the fastest ISDs employing a classical and a quantum computer, respectively;
$\sharp Q$: number of valid $n_0 p \times n_0 p$ block circulant matrices,
$$\sharp Q = \left(\Pi_{i \in \{m_0, \ldots, m_{n_0-1}\}} \binom{p}{i} \right)^{n_0};$$
$\sharp H$: number of valid $p \times n_0 p$ block circulant matrices, $\sharp H = \binom{p}{d_v}^{n_0}$;
FindmPartition(m, n_0): subroutine returning two values. The former one is a sequence of numbers composed as the last integer partition of m in n_0 addends ordered according to the lexicographic order of the reverse sequences, i.e., $\langle m_0, m_1, \ldots, m_{n_0-1} \rangle$, (this allows to get a sequence of numbers as close as possible among them and sorted in decreasing order). The latter returned value is a Boolean value PermanentOk which points out if the partition is legit (**true**) or not (**false**).

1   $p \leftarrow 1$
2   **repeat**
3      $p \leftarrow$ NEXTPRIME(p)
4      $n \leftarrow n_0 p$, $k \leftarrow (n_0 - 1)p$, $r \leftarrow p$
5      $t \leftarrow 1$
6   **while** $\left(t \leq r \wedge \left(\text{C-ISD-COST}(n, k, t) < 2^{\lambda_c} \vee \text{Q-ISD-COST}(n, k, t) < 2^{\lambda_q} \right) \right)$ **do**
7         $t \leftarrow t + 1$
8      $d'_v \leftarrow 4$
9   **repeat**
10        $d_v \leftarrow \lfloor \sqrt{d'_v} \rfloor - 1 - (\lfloor \sqrt{d'_v} \rfloor \bmod 2)$
11   **repeat**
12           $d_v \leftarrow d_v + 2$
13           $m \leftarrow \left\lceil \frac{d'_v}{d_v} \right\rceil$
14           $\langle m_0, m_1, \cdots, m_{n_0-1} \rangle$, PermanentOk $\leftarrow$ FindmPartition(m, n_0)
15   **until** PermanentOk $=$ **true** $\vee$ $(m < n_0)$
16   **if** $(m > n_0)$ **then**
17   SecureOk $\leftarrow$ C-ISD-COST$(n, r, n_0 d'_v) \geq 2^{\lambda_c} \wedge$ Q-ISD-COST$(n, r, n_0 d'_v) \geq 2^{\lambda_q}$
18   SecureOk $\leftarrow$ SecureOk $\wedge$ $\sharp H \geq 2^{\lambda_c} \wedge \sqrt{\sharp H} \geq 2^{\lambda_q} \wedge \sharp Q \geq 2^{\lambda_c} \wedge \sqrt{\sharp Q} \geq 2^{\lambda_q}$
19   **else**
20   SecureOk $\leftarrow$ **false**
21        $d'_v \leftarrow d'_v + 1$
22   **until** (SecureOk $=$ **true** $\vee$ $d'_v n_0 \geq p$)
23   **if** (SecureOk $=$ **true**) **then**
24        $p_{th} \leftarrow$ BF$_{th}(n_0, m d_v, t)$
25   **else**
26        $p_{th} \leftarrow p$
27 **until** $p > p_{th}(1 + \epsilon)$

28 **return** $(p, t, d_v, m, \langle m_0, m_1, \cdots, m_{n_0-1} \rangle)$

Since there are cyclic dependencies among the constraints on p, t, d_v and m, the search for the parameter set is structured as a fixed point solver iterating on a test on the size of p (lines 2–28).

The loop starts by analyzing the next available prime p extracted from a list of pre-computed values such that $\text{ord}_2(p) = p - 1$, and sorted in ascending order (line 3). The length, n, dimension, k, and redundancy, $r = n - k$, of the code are then assigned to obtain a code rate equal to $1 - \frac{1}{n_0}$ (line 4). Subsequently, the procedure for the parameter choice proceeds executing a loop (lines 5–7) to determine a value t, with $t < r$, such that a message recovery attack on a generic code $\mathcal{C}(n, k, t)$ requires more than the specified amount of computational efforts on both classical and quantum computers.

To determine the weight of a column of H, i.e., d_v and the weight of a column of Q, i.e., m, with $m = \sum_{i=0}^{n_0-1} m_i$, the procedure moves on searching for a candidate value of d'_v, where $d'_v = d_v m$ and $d'_v n_0$ is the weight of a row of HQ. Given a value for d'_v (line 8 and line 21), the value of d_v is computed as the smallest odd integer greater than the square root of d'_v (line 10). The condition of d_v being odd is sufficient to guarantee the non singularity of the circulant blocks of H, while the square root computation is meant to distribute the weight d'_v evenly between the weight of a column of H and the weight of a column of Q. The weight of a column of Q, i.e., m, is then computed through the loop in lines 11–15. Specifically, the value of m must allow a partition into n_0 integers (i.e., $m = \sum_{i=0}^{n_0-1} m_i$) such that the permanent of the circulant integer matrix having the said partition as a row is odd, for the matrix Q to be invertible [3]. Therefore, in the loop body the value of m is assumed as $\left\lceil \frac{d'_v}{d_v} \right\rceil$ (line 13) and subsequently checked to derive the mentioned partition in n_0 integers. The loop (lines 11–15) ends when either a valid partition of m is found or m turns out to be smaller than the number of blocks n_0 (as finding a partition in this case would be not possible increasing only the value of d_v).

Algorithm 3 proceeds to test for the security of the cryptosystem against key recovery attacks and key enumeration attacks on both classical and quantum computers (lines 16–18). If a legitimate value for m has not been found, the current parameters of the cryptoystem are deemed insecure (line 20). In line 21, the current value of d'_v is incremented by one and another iteration of the loop is executed if the security constraints are not met with the current parameters (i.e., SecureOk = false) and it is still viable to perform another iteration to check the updated value of d'_v, i.e., $d'_v n_0 < p$ (line 22).

If suitable values for the code parameters from a security standpoint are found, the algorithm computes the minimum value of p, named p_{th}, such that the decoding algorithm is expected to correct t errors, according to the methodology reported in [3] (see lines 23–24); otherwise, the value of p_{th} is forced to be equal to p (lines 25–26) in such a way that another iteration of the outer loop of Algorithm 3 is executed through picking a larger value of p and new values for the remaining parameters.

We note that, while the decoding threshold provides a sensible estimate of the fact that the QC-LDPC code employing the generated parameters will correct the computed amount of errors, this is no substitute for a practical DFR evaluation, which is then performed through Monte Carlo simulations. Willing

Table 2. Parameter sizes for LEDAcrypt KEM obtained with the parameter design tool, compared to those of LEDAkem appearing in the original specification

NIST cat.	n_0	LEDAcrypt KEM					LEDAkem original				
		p	t	d_v	m	Errors out of decodes	p	t	d_v	m	Errors out of decodes
1	2	14,939	136	11	[4,3]	14 out of $1.2 \cdot 10^9$	27,779	224	17	[4,3]	19 out of $2.22 \cdot 10^9$
	3	7,853	86	9	[4,3,2]	0 out of $1 \cdot 10^9$	18,701	141	19	[3,2,2]	0 out of $1 \cdot 10^9$
	4	7,547	69	13	[2,2,2,1]	0 out of $1 \cdot 10^9$	17,027	112	21	[4,1,1,1]	0 out of $1 \cdot 10^9$
3	2	25,693	199	13	[5,3]	2 out of $1 \cdot 10^9$	57,557	349	17	[6,5]	0 out of $1 \cdot 10^8$
	3	16,067	127	11	[4,4,3]	0 out of $1 \cdot 10^9$	41,507	220	19	[3,4,4]	0 out of $1 \cdot 10^8$
	4	14,341	101	15	[3,2,2,2]	0 out of $1 \cdot 10^9$	35,027	175	17	[4,3,3,3]	0 out of $1 \cdot 10^8$
5	2	36,877	267	11	[7,6]	0 out of $1 \cdot 10^9$	99,053	474	19	[7,6]	0 out of $1 \cdot 10^8$
	3	27,437	169	15	[4,4,3]	0 out of $1 \cdot 10^9$	72,019	301	19	[7,4,4]	0 out of $1 \cdot 10^8$
	4	22,691	134	13	[4,3,3,3]	0 out of $1 \cdot 10^9$	60,509	239	23	[4,3,3,3]	0 out of $1 \cdot 10^8$

to target a DFR of 10^{-9}, we enlarged heuristically the value of p until the target DFR was reached (adding 5% of the value of p). Enlargements took place for:

– Category 1: $n_0 = 2$: 6 times, $n_0 = 3$: 1 time, $n_0 = 4$: 1 time
– Category 3: $n_0 = 2$: 4 times, $n_0 = 3$: 0 times, $n_0 = 4$: 0 times
– Category 5: $n_0 = 2$: 0 times, $n_0 = 3$: 0 times, $n_0 = 4$: 0 times.

The C++ tool[3] provided follows the computation logic described in Algorithm 3, but it is optimized to reduce the computational effort as follows:

– The search for the values of t and d'_v respecting the constraints is performed by means of a dichotomic search instead of a linear scan of the range.
– The computations of the binomial coefficients employ a tunable memorized table to avoid repeated re-computation, plus a switch to Stirling's approximation (considering the approximation up to the fourth term of the series) only in the case where the value of $\binom{a}{b}$ is not available in the table and $b > 9$. In case the value of the binomial is not available in the table and $b < 9$ the result is computed with the iterative formula for the binomial, to avoid the discrepancies between Stirling's approximation and the actual value for small values of b.
– The values of p respecting the constraint $\text{ord}_2(p) = p - 1$ are pre-computed up to $159,979$ and stored in a lookup table.
– The search for the value of p is not performed scanning linearly the aforementioned table. The strategy to find the desired p starts by setting the value of the candidate for the next iteration to NEXTPRIME($\lceil (1 + \epsilon)p_{th} \rceil$) up to

---

[3] The C++ tool relies on Victor Shoup's NTL library (available at https://www.shoup.net/ntl/), in particular for the arbitrary precision integer computations and the tunable precision floating point computations, and requires a compiler supporting the C++11 standard.

Table 3. Computational cost of an exhaustive enumeration attack on either the matrix H or the matrix Q. The quantum execution model considers the possibility of attaining the full speedup yielded by the application of Grover's algorithm to the computation

NIST cat.	n_0	H enumeration cost ($\log_2$ ♯binary op.s)		Q enumeration cost ($\log_2$ ♯quantum gates)	
		Classical	Quantum	Classical	Quantum
1	2	254.55	127.27	179.79	89.89
	3	295.93	147.96	326.84	163.42
	4	539.64	269.82	348.68	174.34
3	2	315.79	157.89	247.34	123.67
	3	385.30	192.65	425.80	212.90
	4	667.42	333.71	474.74	237.37
5	2	283.24	145.62	350.84	175.42
	3	542.70	271.35	451.27	225.63
	4	622.26	311.13	703.06	351.53

finding a value of p, $\bar{p}$, which satisfies the constraints. Subsequently the algorithm starts scanning the list of primes linearly from $\bar{p}$ backwards to find the smallest prime which satisfies the constraints.

6   LEDAcrypt Instances with Ephemeral Keys

In Table 2 we provide parameters for LEDAcrypt KEM instances employing ephemeral keys, and compare them with those of LEDAkem appearing in the original specification. This shows how the new parameterization is tighter and enables a considerable reduction in the key sizes. Deriving them took approximately a day for all the parameter sets with $n_0 \in \{3, 4\}$ and approximately a month for all the parameter sets with $n_0 = 2$ on a dual socket AMD EPYC 7551 32-Core CPU. The memory footprint for each parameter seeking process was below 100 MiB.

6.1   Resulting Computational Complexity of Attacks

When an algorithmic procedure is exploited for the design of parameter sets, as in our case, some constraints on the choice of the row/column weights of H and Q must be imposed in such a way as to make enumeration of either H or Q unfeasible to an attacker. Therefore, enumeration attacks of the type described in Sect. 4.1 must be taken into account. In Table 3 we report the computational cost of performing such an exhaustive enumeration, both with a classical and a quantum computer. The latter has been obtained by applying the speedup due to Grover's algorithm to the complexity computed considering a classical computer.

Table 4. Cost of performing a message recovery attack, i.e., an ISD on the code $\mathcal{C}(n_0 p, (n_0 - 1)p, t)$, for LEDAcrypt instances with the parameters p, t reported in Table 2, employing the considered ISD variants.

NIST cat.	n_0	Classical computer (log$_2$ ♯binary op.s)						Quantum computer (log$_2$ ♯quantum gates)	
		Prange [34]	L-B [24]	Leon [25]	Stern [36]	F-S [11]	BJMM [5]	Q-LB [8]	Q-Stern [8]
1	2	169.05	158.23	156.35	148.59	148.57	144.37	97.26	98.67
	3	167.72	157.37	154.51	147.67	147.65	144.29	96.14	97.55
	4	169.62	159.40	155.86	149.32	149.31	145.98	97.47	98.22
3	2	234.11	222.19	220.26	210.42	210.41	207.17	130.22	131.62
	3	235.32	223.84	220.82	211.91	211.90	208.71	130.67	132.07
	4	235.98	224.66	220.97	212.39	212.39	209.10	131.26	132.66
5	2	303.56	290.79	288.84	277.40	277.39	274.54	165.18	166.58
	3	303.84	291.53	288.42	277.98	277.98	274.34	165.48	166.88
	4	303.68	291.54	287.78	277.67	277.67	274.91	165.52	166.92

From the results in Table 3 it is straightforward to note that an exhaustive search on either H or Q is above the required computational effort.

As described in Sect. 4.2, the two main attacks that can be mounted against the considered systems are message recovery attacks and key recovery attacks based on ISD algorithms. Tables 4 and 5 report the complexities of these attacks against LEDAcrypt instances employing the parameters p, t in Table 2. An interesting point to be noted is that, while providing clear asymptotic speedups, the improvements to the ISD algorithms proposed since Stern's [36] are only able to achieve a speedup between 2^2 and 2^4 when their finite regime complexities are considered in the range of values concerning LEDAcrypt cryptosystem parameters. Concerning quantum ISDs, it is interesting to notice that the quantum variant of the Stern's algorithm as described by de Vries [8] does not achieve an effective speedup when compared against a quantum transposition of Lee and Brickell's ISD. Such a result can be ascribed to the fact that the reduction in the number of ISD iterations which can be obtained by Stern's ISD is mitigated by the fact that the applying Grover's algorithm to the iterations themselves cuts their number (and Stern's reduction factor) quadratically [7].

Comparing the computational complexities of the message recovery attack (Table 4) and the key recovery attack (Table 5), we note that performing a message recovery attack is almost always easier than the corresponding key recovery attack on the same parameter set, albeit by a small margin.

7   LEDAcrypt Instances with Long Term Keys

For LEDAcrypt instances employing long term keys, we need that the DFR is sufficiently small to enable IND-CCA2. However, such small values of DFR cannot be assessed through Monte Carlo simulations. Hence, for these instances we consider a Q-decoder performing two iterations and exploit the analysis reported

Table 5. Cost of performing a key recovery attack, i.e., an ISD on the code $\mathcal{C}(n_0 p, p, n_0 d_v m)$, for the revised values of the parameters p, n_0, d_v, m reported in Table 2, employing the considered ISD variants

NIST cat.	n_0	Classical computer (log$_2$ ♯binary op.s)						Quantum computer (log$_2$ ♯quantum gates)	
		Prange [34]	L-B [24]	Leon [25]	Stern [36]	F-S [11]	BJMM [5]	Q-LB [8]	Q-Stern [8]
1	2	180.25	169.06	167.24	158.76	158.75	154.94	99.21	100.62
	3	169.36	157.78	156.53	147.71	147.68	144.08	93.93	95.34
	4	179.86	167.79	165.69	157.13	157.10	153.01	99.71	101.12
3	2	237.85	225.77	223.87	213.72	213.71	210.64	128.35	129.75
	3	241.70	228.98	227.03	216.59	216.57	213.18	130.56	131.96
	4	254.92	241.73	238.97	228.80	228.78	224.76	137.60	139.01
5	2	315.08	302.11	300.19	288.35	288.34	285.71	167.04	168.44
	3	320.55	306.93	304.48	292.78	292.77	289.00	170.31	171.71
	4	312.68	298.84	295.66	284.59	284.58	280.91	166.82	168.22

in Appendix A in order to characterize its DFR. To this end, we consider a two-stage rejection sampling during key generation, that is:

1. Only pairs of H and Q such that the weight of a column of $H' = HQ$ is $d_v m$.
2. Only pairs of H and Q are retained for which the condition (2) in Appendix A.1 guaranteeing low enough DFR is verified.

The first rejection sampling is useful to achieve a constant computational effort for message and key recovery attacks. In addition, it simplifies the second rejection sampling, which can take advantage of a special case of the analysis reported in Appendix A. Then, the second rejection sampling is performed to ensure that the generated key pair can achieve with two iterations of the Q-decoder the correction of all residual errors of weight $\leq \bar{t}$ left by the first decoder iteration. Such a property can be obtained with the knowledge of the generated matrices H and Q alone. If the above condition is not satisfied the generated key pair is discarded, and the procedure is repeated until a valid key pair is found. For valid key pairs, achieving a desired target DFR value, $\overline{\text{DFR}}$, is guaranteed by the choice of code parameters such that the first iteration of the Q-decoder results in at most $\bar{t}$ residual errors with probability $> 1 - \overline{\text{DFR}}$.

Given a chosen target DFR and a set of parameters for a LEDAcrypt instance, we are able to evaluate the amount of secret keys which allow achieving the desired DFR target. Such a procedure is integrated in the key generation process for LEDAcrypt instances with long term keys, where the concern on the DFR is actually meaningful, as opposed to instances with ephemeral key pairs.

Some choices are reported in Table 6, by imposing a DFR bounded by either 2^{-64} or $2^{-\lambda}$, where λ equals 128, 192, 256 for NIST Category 1, 3, 5, respectively. The proposed choices aim at parameter sets for which the probability of drawing a random secret key achieving the DFR target is significant to minimize key generation overhead. To design these parameters, we start from the ones obtained

Table 6. Parameters for the LEDAcrypt KEM with long term keys and the LEDAcrypt PKC employing a two-iteration Q-decoder matching a DFR equal to 2^{-64} and a DFR equal to $2^{-\lambda}$, where λ equals 128, 192, 256 for NIST Category 1, 3, 5, respectively

NIST category	n_0	DFR	p	t	d_v	m	$\bar{t}$	No. of keys out of 100 with the required DFR	b_0
1	2	2^{-64}	35,899	136	9	[5,4]	4	95	44
	2	2^{-128}	52,147	136	9	[5,4]	4	95	43
3	2	2^{-64}	57,899	199	11	[6,5]	5	92	64
	2	2^{-192}	96,221	199	11	[6,5]	5	92	64
5	2	2^{-64}	89,051	267	13	[7,6]	6	93	89
	2	2^{-256}	152,267	267	13	[7,6]	6	93	88

through the automated parameter optimization procedure used for instances with ephemeral keys previously described. Then, we proceed by increasing the size of the circulant blocks while keeping the product md_v constant or slightly increased, until we obtain a probability larger than the given target that the number of bit errors that are left uncorrected by the first iteration is $\leq \bar{t}$. In particular, such a probability is computed by considering all possible choices for the flipping threshold of the first iteration, and by taking the optimal one (i.e., the one corresponding to the maximum value of the probability). Note that such changes may only impact positively on the security margin against ISD attacks and key enumeration attacks.

For any set of parameters so designed, we draw 100 key pairs at random, and evaluate how many of them satisfy the condition (2) in Appendix A. As it can be seen from the results reported in Table 6, the parameter sets we determined are able to achieve a DFR $< 2^{-64}$ increasing the key size by a factor ranging from 2× to 3× with respect to the case of ephemeral key pairs.

The obtained LEDAcrypt parameterizations show that it is possible to achieve the desired DFR discarding an acceptable number of key pairs, given proper tuning of the parameters. The parameter derivation procedure for these LEDAcrypt instances can also be automated, which could be advantageous in terms of flexibility in finding optimal parameters for a given code size or key rejection rate.

8 Conclusion

In this work we presented a threefold contribution on code-based cryptosystems relying on QC-LDPC codes, such as the NIST post-quantum candidate LEDAcrypt. First of all, we quantify the computational effort required to break AES both via classical and quantum computation, providing a computational effort to be matched in post-quantum cryptosystem parameter design. Our second contribution is an automated optimization procedure for the parameter generation of ephemeral-key cryptosystem parameters for LEDAcrypt, providing a 10^{-9}–10^{-8} DFR, low enough for practical use. Our third contribution is a closed form characterization of the decoding strategy employed in the LEDAcrypt systems that allows to obtain an upper bound on their DFR at design time, in turn the generation of cryptosystem parameter sets which guarantee a bounded DFR. This in turn allows to generate cryptosystem parameters achieving IND-CCA2 guarantees, which ensure that active attacks, including reaction attacks, have a computational cost exceeding the desired security margin, without the need for ephemeral keys. We note that the proposed bound on the DFR can be fruitfully integrated in the automated design procedure proposed in this paper.

A   Bounded DFR for Q-Decoders

Binary block error correction codes $\mathcal{C}(n, k, t)$ with a low-density $r \times n$ parity-check matrix H' allow iterative decoding strategies which aim at solving at fixed point the simultaneous binary equation system given by $s = H'e^T$, where $s \in \mathbb{Z}_2^r$ is a $1 \times r$ binary vector named *syndrome*, $e \in \mathbb{Z}_2^n$ is a $1 \times n$ binary vector with a given number $t \ll n$ of non-zero entries named error vector, representing the unknown sequence of values to be found, while H' is assumed to have $d_v \ll n$ non-zero entries per column. Therefore, the purpose of an iterative decoding procedure is to compute the values of the elements of e given H' and s.

A common approach to perform iterative decoding is the *Bit Flipping* (BF) strategy firstly described in [13]. Such an approach considers the i-th row of H', with $i \in \{0, \ldots, r-1\}$, as a representation of the coefficients of the parity check equation involving the unknown e_j, with $j \in \{0, \ldots, n-1\}$, having as constant term the i-th element of the syndrome s. Each coefficient is associated to a binary variable $e_j \in \mathbb{Z}_2$, i.e., a binary element of the error vector e whose value should be determined. Initially, the guessed value of the error vector, denoted in the following as $\hat{e}$, is assumed to be the null vector, i.e., $\hat{e} = 0_{1 \times n}$ (i.e., the bits of the received message are initially assumed to be all uncorrupted).

The iterative BF decoding procedure repeats (at least one time) the execution of two phases (named in the following as *Count of the unsatisfied parity checks*, and *Bit-flipping*, respectively) until either all the values of the syndrome become null (pointing out the fact that every value of e has been found) or an imposed a-priori maximum number of iterations, $l_{\max} \geq 1$, is reached.

1. *Count of the unsatisfied parity checks.* The first phase of the decoding procedure analyzes the parity check equations where a given error variable $\hat{e}_j$ is involved, with $j \in \{0, \ldots, n-1\}$, i.e., the number of rows of H' where the j-th element is non-zero, and counts how many of them are unsatisfied. In other words, it counts how many equations involving the unknown e_j have a constant term in the syndrome which is non-zero. Such a count of the number of unsatisfied parity check equations, upc_j, can be computed for each error variable $\hat{e}_j$, lifting the elements of s and H' from $\mathbb{Z}_2$ to $\mathbb{Z}$ and performing a product between an integer vector ($\varsigma \leftarrow \text{Lift}(s)$) by an integer matrix ($\mathcal{H}' \leftarrow \text{Lift}(H')$), obtaining a $1 \times n$ integer vector $\text{upc}^{(\text{BF})}$, i.e., $\text{upc}^{(\text{BF})} \leftarrow \varsigma \, \mathcal{H}'$.

2. *Bit-flipping.* The second phase changes (i.e., flips, hence the name bit-flipping) each value of an error variable $\hat{e}_j$ for which $\mathtt{upc}_j^{(\mathrm{BF})}$ exceeds a given threshold $b \geq 1$. Subsequently, it updates the value of the syndrome, computing it as $H'\hat{e}^T$, employing the new value of the $\hat{e}_j$ variables in the process.

The LEDA cryptosystems leverage Q-decoders that achieve smaller complexity than classical BF decoders. In fact, due to the sparsity of both H and Q, their product HQ has a number of non-zero row elements $\leq d_c\, m$, with the equality holding with very high probability. Such a fact can be exploited to perform the first phase of the bit-flipping decoding procedure in a more efficient way. To do so, the Q-decoder proceeds to lift H and Q in the integer domain obtaining $\mathcal{H} \leftarrow \mathtt{Lift}(H)$ and $\mathcal{Q} \leftarrow \mathtt{Lift}(Q)$, respectively. Subsequently, it performs a decoding strategy similar to the one described above, as follows.

1. *Count of the unsatisfied parity checks.* The first phase is performed in two steps. First of all, a temporary $1 \times n$ vector of integers $\mathtt{upc}^{(\mathrm{temp})}$ is computed in the same fashion as in the BF decoder, employing the lifted syndrome, $\varsigma \leftarrow \mathtt{Lift}(s)$, and $\mathcal{H}$ instead of $\mathcal{H}'$, i.e., $\mathtt{upc}^{(\mathrm{temp})} \leftarrow \varsigma\, \mathcal{H}$. The value of the actual $1 \times n$ integer vector $\mathtt{upc}^{(\mathrm{Q-dec})}$ storing the unsatisfied parity-check counts is then computed as: $\mathtt{upc}^{(\mathrm{Q-dec})} \leftarrow \mathtt{upc}^{(\mathrm{temp})}\, \mathcal{Q}$.
2. *Bit-flipping.* The second phase of the Q-decoder follows the same steps of the BF one, flipping the values of the guessed error vector $\hat{e}_j, j \in \{0, \ldots, n-1\}$, for which the j-th unsatisfied parity-check count $\mathtt{upc}_j^{(\mathrm{Q-dec})}$ exceeds the chosen threshold b. Subsequently, the value of the syndrome s is updated as $s+HQ\hat{e}^T$.

In both the BF- and Q-decoder, the update to the syndrome value caused by the flipping of the values of $\hat{e}$ in the second phase of the procedure, can be computed incrementally, adding only the contributions due to the value change of $\hat{e}$ (see the LEDA cryptosystems specification [3]).

If a null s is obtained before the maximum allowed number of iterations $l_{\max}$ is exceeded, then the Q-decoder terminates with success its decoding procedure, otherwise it ends with a decoding failure.

Lemma 1 (Equivalence of the bit-flipping decoder and Q-decoder).
Let H and Q be the two matrices composing the parity-check matrix $H' = HQ$, and denote as $\mathcal{H}' \leftarrow \mathtt{Lift}(H'), \mathcal{H} \leftarrow \mathtt{Lift}(H), \mathcal{Q} \leftarrow \mathtt{Lift}(Q)$, the matrices obtained through lifting their values from $\mathbb{Z}_2$ to $\mathbb{Z}$. Assume a BF procedure acting on H' and a Q-decoding procedure acting on $\mathcal{H}$ and $\mathcal{Q}$, both taking as input the same syndrome value s, providing as output an updated syndrome and a guessed error vector $\hat{e}$ (which is initialized as $\hat{e} = 0_{1 \times n}$ at the beginning of the computations), and employing the same bit-flipping thresholds. If $\mathcal{H}' = \mathcal{H}\mathcal{Q}$, the BF and Q-decoding procedures compute as output the same values for s and $\hat{e}$.

Proof. The functional equivalence can be proven showing that the update to the two state vectors, the syndrome s and the current $\hat{e}$, performed by the bit-flipping decoder and the Q-decoder leads to the same values at the end of each iteration of the decoding algorithms. We start by observing that the second phase of the BF and Q-decoder procedure will lead to the same state update of s and $\hat{e}$ if the values of the $\mathbf{upc}^{(\text{BF})}$ vector for the BF procedure and the $\mathbf{upc}^{(\text{Q-dec})}$ vector for the Q-decoder coincide. Indeed, since the update only depends on the values of the unsatisfied parity-checks and the flipping threshold b, if $\mathbf{upc}^{(\text{BF})} = \mathbf{upc}^{(\text{Q-dec})}$ the update on $\hat{e}$ and s will match. We consider, from now on, the parity-check computation procedures as described before through matrix multiplications over the integer domain, and prove that, during the first phase, the BF decoder and the Q-decoder yield values of $\mathbf{upc}^{(\text{BF})}$ and $\mathbf{upc}^{(\text{Q-dec})}$ such that $\mathbf{upc}^{(\text{BF})} = \mathbf{upc}^{(\text{Q-dec})}$ under the hypothesis that the starting values for s and $\hat{e}$ match. Considering the computation of $\mathbf{upc}^{(\text{BF})}$, and denoting with h'_{ij} the element of H' at row i, column j, we have that $\mathbf{upc}^{(\text{BF})} =$

$\varsigma \mathcal{H}'$, hence $\mathbf{upc}_j^{(\text{BF})} = \sum_{z=0}^{r-1} h'_{zj} s_z$. The computation of $\mathbf{upc}^{(\text{Q-dec})}$ proceeds as

follows $\mathbf{upc}^{(\text{Q-dec})} = (\varsigma \mathcal{H}) \mathcal{Q} = \sum_{i=0}^{n-1} \left(\sum_{z=0}^{r-1} s_z h_{zi} \right) q_{ij} = \sum_{z=0}^{r-1} \left(\sum_{i=0}^{n-1} h_{zi} q_{ij} \right) s_z.$

Recalling the hypothesis $\mathcal{H}' = \mathcal{HQ}$, it is possible to acknowledge that

$$\sum_{i=0}^{n-1} h_{zi} q_{ij} = h'_{zj},$$ which, in turn, implies that $\mathbf{upc}^{(\text{Q-dec})} = \mathbf{upc}^{(\text{BF})}$.   □

Lemma 2 (Computational advantage of the Q-decoder). *Let us consider a bit-flipping decoding procedure and a Q-decoder procedure both acting on the same parity matrix $H' = HQ$. The number of non-zero entries of a column of H is $d_v \ll n$, the number of non-zero entries of a column of Q is $m \ll n$, and the number of non-zero entries of a column of H' is $d_v m$ (assuming no cancellations occur in the multiplication HQ). The computational complexity of an iteration of the bit-flipping decoder equals $\mathcal{O}(d_v mn + n)$, while the computational complexity of an iteration of the Q-decoder procedure is $\mathcal{O}((d_v + m)n + n)$.*

Proof (Sketch). The proof can be obtained in a straightforward fashion with a counting argument on the number of operations performed during the iteration of the decoding procedures, assuming a sparse representation of H, H' and Q. In particular the amount of operations performed during the unsatisfied parity-check count estimation phase amounts to $\mathcal{O}(d_v mn)$ additions for the bit-flipping decoder and to $\mathcal{O}((d_v + m)n)$ for the Q-decoder, while both algorithms will perform the same amount of bit flips $\mathcal{O}(n + r) = \mathcal{O}(n)$ in the bit-flipping and syndrome update computations.   □

If $\mathcal{H}' \neq \mathcal{HQ}$, it is not possible to state the equivalence of the two procedures. However, some qualitative considerations about their behavior can be drawn analyzing the product $\mathcal{HQ}$ in the aforementioned case. Indeed, an entry in the i-th row, j-th column of $\mathcal{H}'$ is different from the one with the same coordinates in $\mathcal{HQ}$ whenever the scalar product of the i-th row of $\mathcal{H}$ and the j-th column of $\mathcal{Q}$ is ≥ 2. First of all, we note that such an event occurs with probability $\sum_{i=2}^{\min\{m,d_c\}} \frac{\binom{m}{i}\binom{n-m}{d_c-i}}{\binom{n}{d_c}}$, which becomes significantly small if the code parameters

(n, d_c, m) take values of practical interest. Since the number of entries which have a different value in $\mathcal{H}\mathcal{Q}$ with respect to $\mathcal{H}'$ are expected to be small, the values of the unsatisfied parity check counts $\mathbf{upc}^{(\mathrm{BF})}$ and $\mathbf{upc}^{(\mathrm{Q-dec})}$ are also expected to differ by a quite small amount, while the computational complexity of the decoding algorithms will remain substantially unchanged with respect to the case in which $\mathcal{H}' = \mathcal{H}\mathcal{Q}$, as the term $d_v m$ in the BF decoder is reduced by a significantly small term, while the one of the Q-decoder is unchanged.

The decoding failure rate of the Q-decoder is crucially dependent on the choice made for the bit-flipping threshold b. Indeed, the designer aims at picking a value of b satisfying the following criteria.

(i) The number of values of the unsatisfied parity-check count $\mathbf{upc}_j$, with $j \in \{0, \ldots, n-1\}$ such that $\hat{e}_j \neq e_j$ and $\mathbf{upc}_j \geq b$ should be maximum. Indeed, when this situation occurs, all such values $\mathbf{upc}_j$ are rightfully flipped.

(ii) The number of values of the unsatisfied parity-check count $\mathbf{upc}_j$, with $j \in \{0, \ldots, n-1\}$ such that $\hat{e}_j = e_j$ and $\mathbf{upc}_j < b$ should be maximum. Indeed, when this situation occurs, all such values $\mathbf{upc}_j$ are rightfully not flipped.

Observing that during the decoding procedure the j-th bit value of the guessed error vector $\hat{e}$ is flipped when $\mathbf{upc}_j$ is higher than or equal to b, an ideal case where it is possible to attain a null DFR is the one in which, whenever the maximum possible value of any unsatisfied parity-check count $\mathbf{upc}_j$ related to a variable $\hat{e}_j = e_j$ (i.e., no flip is needed) is lower than the minimum possible value of $\mathbf{upc}_k$ related to any variable $\hat{e}_k \neq e_k$ (i.e., flip is needed). Indeed, in this case, setting the threshold to any value b such that $\mathtt{max\_upc}_{no\,flip} < b \leq \mathtt{min\_upc}_{flip}$ allows the Q-decoder to compute the value of the actual error vector e in a single iteration.

To provide code parameter design criteria to attain a zero DFR in a single iteration with the Q-decoder, we now analyze the contribution to the values of $\mathbf{upc}$ provided by the bits of the actual error vector. Let $u_z \in \mathbb{Z}_2^n$, with $z \in \{0, \ldots, n-1\}$, denote $1 \times n$ binary vector such that only the z-th component of u_z is non-zero (i.e., it has unitary Hamming weight, $wt(u_z) = 1$). We now consider the actual error vector e as the sum of $t \geq 1$ vectors $\in \mathbf{U}(e) = \{u_z \in \mathbb{Z}_2^n, \; wt(u_z) = 1, z \in \mathbf{I}(e)\}$, where $\mathbf{I}(e) \subset \{0, \ldots, n-1\}$ defines the support of e and $|\mathbf{I}(e)| = t$ (thus it also holds $|\mathbf{U}(e)| = t$), and quantify the contributions of each bit in e to the value of $\mathbf{upc}_z$ computed by the Q-decoder in its first iteration, proceeding backwards from each $u_z \in \mathbf{U}(e)$ composing e.

We describe the mentioned quantification with the aid of a running example referred to the syndrome Q-decoding procedure of a toy code $\mathcal{C}(5, 3, 2)$, assuming that a single bit in the actual error vector is asserted, i.e., $e = u_2$. Figure 2 reports a graphical description of the mentioned running example. Our aim is to define a $d_v m \times n$ matrix $P_{(z)}$ containing a set of parity-check equations, i.e., rows of H which contribute to $\mathbf{upc}_z$, $z \in \{0, \ldots, n-1\}$[4].

---

[4] Note that the notation $P_{(z)}$ denotes a matrix whose values are related with the bit in position z of the actual (unknown) error vector (it may include repeated rows).

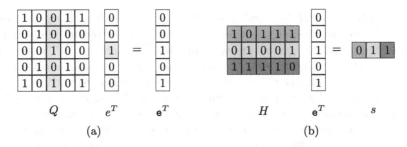

Fig. 2. Steps of the syndrome computation process of a toy code $\mathcal{C}(5,3,2)$, having H with constant column weight $d_v = 2$ and Q with constant column weight $m = 2$. The single bit error vector employed is $e = u_2 = (0,0,1,0,0)$. In (a) the error vector is expanded as $\mathbf{e} = u_2 + u_4 = (0,0,1,0,1)$ after the multiplication by Q, $\mathbf{e} = (Qe^T)^T$. In (b) the effect on the syndrome of multiplying $\mathbf{e}$ by H is shown, i.e., $s = (H(Qe^T))^T$. (Color figure online)

Consider the syndrome value $s = (H(Qe^T))^T$ obtained as the multiplication between the matrix Q and the actual error vector value $e = u_z$, with $z \in \{0, \ldots, n-1\}$, (i.e., $Q\,e^T = Q\,u_z^T$), followed by the multiplication between the matrix H and the *expanded error* vector $\mathbf{e}^{(z)} = Q\,e^T$, with $wt(\mathbf{e}^{(z)}) = m$, which has been computed in the previous step (i.e., $s = H\left((\mathbf{e}^{(z)})^T\right)^T$. Note that, in this case $\mathbf{e}^{(z)}$ is the result of a computation depending only on the value of Q and z, not on the actual error vector being decoded.

Consider $\mathbf{e}^{(z)}$ as the sum of m binary vectors u_j, $j \in \{0, \ldots, n-1\}$ with a single non-zero entry in the j-th position (see $\mathbf{e}^{(2)} = u_2 + u_4$, with u_2 highlighted in red and u_4 in blue in Fig. 2). Each u_j in $\mathbf{e}^{(z)}$ is involved in all the parity-check equations of H, i.e., the $1 \times n$ rows, H_i, with $i \in \{0, \ldots, r-1\}$, having their j-th element set to 1. It is thus possible to build a $d_v m \times n$ matrix $P_{(z)}$ juxtaposing all the parity-check equations in H such that their j-th element is set to 1, for all the u_j composing $\mathbf{e}^{(z)}$. The $P_{(z)}$ matrix allows to compute the contribution to the value $\mathbf{upc}_z$, $z \in \{0, \ldots, n-1\}$ provided by any expanded error vector $\mathbf{e}^{(j)}$, $j \in \{0, \ldots, n-1\}$, as it is constituted by all and only the parity-check equations which will have their non null results counted to obtain $\mathbf{upc}_z$. Therefore, for any binary variable e_z, $z \in \{0, \ldots, n-1\}$, in the actual error vector e, it is possible to express the value of the corresponding unsatisfied parity-check count evaluated by the decoding procedure as $\mathbf{upc}_z \leftarrow wt\left(P_{(z)}\left(\mathbf{e}^{(z)}\right)^T\right)$.

The construction of $P_{(z)}$ for the toy example is reported in Fig. 3, where $z = 2$. $P_{(z)}$ is obtained by juxtaposing the rows H_0 and H_2, as they both involve u_2, and juxtaposing to them the rows H_0 and H_1 as they both involve u_4.

Figure 3 provides a visual comparison of the two equivalent processes to compute the value of $\mathbf{upc}_2$ considering the values of ς, $\mathcal{H}$ and $\mathcal{Q}$ obtained as the integer

---

The round brackets employed in the subscript are meant to disambiguate this object from the notations related to the z-th row of a generic matrix P, i.e., P_z.

lifts of s, H and Q from Fig. 2. Indeed, computing the value of $\mathbf{upc}_2$ as the third component of the vector $\varsigma \mathcal{H} \mathcal{Q}$ yields the same results as computing the binary vector $P_{(2)}(\mathbf{e}^{(2)})^T$ and computing its weight.

Relying on the $P_{(z)}$ matrices to express the contribution of a given expanded error vector $\mathbf{e}^{(z)}$, we are able to rewrite the computation of $\mathbf{upc}_z$ for a generic error vector with t asserted bits, i.e., $e = \sum\limits_{u_z \in \mathbf{U}(e)} u_z$, where $\mathbf{I}(e) \subset \{0, \ldots, n-1\}$ with $|\mathbf{I}(e)| = t$, and $\mathbf{U}(e) = \{u_i, \ wt(u_i) = 1, \ i \in \mathbf{I}(e)\}$, as follows

$$\mathbf{upc}_z \leftarrow wt \left(\sum_{i \in \mathbf{I}(e) \subset \{0,\ldots,n-1\}} P_{(z)} \left(\mathbf{e}^{(i)} \right)^T \right) = wt \left(\sum_{u_z \in \mathbf{U}(e)} P_{(z)} (Q \, u_z^T) \right).$$

A.1   Q-Decoders with Zero DFR

Having provided a way to derive the contribution of any bit of an actual error vector to a specific unsatisfied parity-check count $\mathbf{upc}_z$, following the work in [37] for the BF-decoder, we now proceed to analyze the case of a Q-decoder which is always able to correct all the errors in a single iteration of the decoding procedure. To do so, the bit-flipping action should flip the value of all the elements of the guessed error $\hat{e}$ which do not match e. Recalling that $\hat{e}$ is initialized to the null vector, the first iteration of the Q-decoder should thus flip all the elements $\hat{e}_z$ such that $e_z = 1$.

The Q-decoder will perform all and only the appropriate flips if the $\mathbf{upc}_z$ with $z \in \{0, \ldots, n-1\}$ such that $e_z = 1$, match or exceed the threshold b, and all the $\mathbf{upc}_z$ such that $e_z = 0$ are below the same flipping threshold.

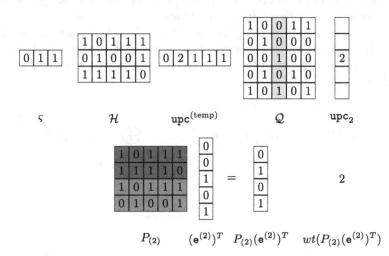

Fig. 3. Representation of the $P_{(z)}$ matrix for the running example ($z = 2$), and computation of the $\mathbf{upc}_z$ value both via $\mathcal{Q}$ and $\mathcal{H}$, and employing $P_{(z)}$

We have that, if the highest value of upc_z when $e_z = 0$ (i.e., $\mathrm{max\_upc}_{no\,flip}$) is smaller than the lowest value of upc_z when $e_z = 1$ (i.e., $\mathrm{min\_upc}_{flip}$), the Q-decoder will be able to correct all the errors in a single iteration if the bit-flipping threshold b is set to $b = \mathrm{min\_upc}_{flip}$, as this will cause the flipping of all and only the incorrectly estimated bits in the guessed error vector $\hat{e}$.

In the following, we derive an upper bound on the maximum admissible number of errors t which guarantees that $\mathrm{max\_upc}_{no\,flip} < \mathrm{min\_upc}_{flip}$ for a given code.

Theorem 1. *Let H be an $r \times n$ parity-check matrix, with constant column weight equal to d_v, and let Q be an $n \times n$ matrix with constant row weight equal to m. Let e be a $1 \times n$ binary error vector with weight t, composed as $e = \sum_{i \in \mathbf{I}(e)} u_i$, with $\mathbf{I}(e) \subset \{0, \ldots, n-1\}$ defining the support of e, $|\mathbf{I}(e)| = t$, where $u_i \in \mathbb{Z}_2^n$, and $wt(u_i) = 1$; and let $\mathbf{e}$ be the $1 \times n$ binary vector computed as $\mathbf{e} = \left(Q e^T\right)^T$. The first iteration of the Q-decoder, taking as input the $1 \times r$ syndrome $s = \left(H e^T\right)^T$, retrieves the values of all the bits of actual error vector e if $t < \frac{\alpha + \beta}{\gamma + \beta}$, where*

$$\begin{cases} \alpha = \min_{z \in \mathbf{I}(e)} \left\{ wt\left(P_{(z)}\left(\mathbf{e}^{(z)}\right)^T\right)\right\}, \\ \beta = \max_{z, i \in \mathbf{I}(e), z \neq i} \left\{ wt\left(P_{(z)}\left(\mathbf{e}^{(z)}\right)^T \wedge P_{(z)}\left(\mathbf{e}^{(i)}\right)^T\right)\right\}, \\ \gamma = \max_{z, i \in \mathbf{I}(e), z \neq i} \left\{ wt\left(P_{(z)}\left(\mathbf{e}^{(i)}\right)^T\right)\right\}, \end{cases} \tag{1}$$

*where $\wedge$ indicates the component-wise binary product (i.e., the Boolean **and**).*

Proof. We first determine the lower bound $\mathrm{min\_upc}_{flip}$ as the lowest value of an unsatisfied parity-check count upc_z, $z \in \{0, \ldots, n-1\}$, when the value of the corresponding bit in the actual error vector is set, i.e., $e_z = 1$.

We start by decomposing the value of upc_z into the contributions provided by the t vectors $u_i \in \mathbb{Z}_2^n$, with $i \in \{0, \ldots, n-1\}$ and $wt(u_i) = 1$, i.e., $e = \sum_{i \in \mathbf{I}(e)} u_i$, $\mathbf{I}(e) \subset \{0, \ldots, n-1\}, |\mathbf{I}(e)| = t$ (thus, $z \in \mathbf{I}(e)$). In other words, we want to quantify the minimum value for the count of the unsatisfied parity checks related to a bit of the guessed error vector that needs to be flipped when it is right to do so. We then have

$$\mathrm{upc}_z = wt\left(P_{(z)}\,\mathbf{e}^T\right) = wt\left(P_{(z)}\left(\mathbf{e}^{(z)}\right)^T \oplus \bigoplus_{\substack{\mathbf{e}^{(i)} = (Q u_i^T)^T \\ u_i \in \mathbf{U}(e) \setminus \{u_z\}}} P_{(z)}\left(\mathbf{e}^{(i)}\right)^T\right).$$

Considering that, for a generic pair of binary vectors a, b of length n we have that $wt(a \oplus b) = wt(a) + wt(b) - 2wt(a \wedge b)$, we expand the former onto

$$\mathtt{upc}_z = wt\left(P_{(z)}\left(\mathbf{e}^{(z)}\right)^T\right) + wt\left(\bigoplus_{\substack{\mathbf{e}^{(i)}=(Qu_i^T)^T \\ u_i \in \mathbf{U}(e)\backslash\{u_z\}}} P_{(z)}\left(\mathbf{e}^{(i)}\right)^T\right) +$$

$$- 2\, wt\left(P_{(z)}\left(\mathbf{e}^{(z)}\right)^T \wedge \bigoplus_{\substack{\mathbf{e}^{(i)}=(Qu_i^T)^T \\ u_i \in \mathbf{U}(e)\backslash\{u_z\}}} P_{(z)}\left(\mathbf{e}^{(i)}\right)^T\right).$$

Recalling that, for two binary vectors a, b it holds that $wt(b) \geq wt(a \wedge b)$, it is worth noting that it also holds that $wt(a)+wt(b)-2\,wt(a\wedge b) \geq wt(a)-wt(a\wedge b)$. Therefore, considering the second and third addend of the above equality on $\mathtt{upc}_z$, we obtain

$$\mathtt{upc}_z \geq wt\left(P_{(z)}\left(\mathbf{e}^{(z)}\right)^T\right) - wt\left(P_{(z)}\left(\mathbf{e}^{(z)}\right)^T \wedge \bigoplus_{\substack{\mathbf{e}^{(i)}=(Qu_i^T)^T \\ u_i \in \mathbf{U}(e)\backslash\{u_z\}}} P_{(z)}\left(\mathbf{e}^{(i)}\right)^T\right).$$

Since we are interested in quantifying the lower bound $\mathtt{min\_upc}_{flip}$ for $\mathtt{upc}_z$, a straightforward rewriting of the previous inequality is as follows

$$\mathtt{upc}_z \geq \min_{z \in \mathbf{I}(e)}\left\{wt\left(P_{(z)}\left(\mathbf{e}^{(z)}\right)^T\right)\right\} +$$

$$- \max_{\substack{z,i \in \mathbf{I}(e) \\ z \neq i}}\left\{wt\left(P_{(z)}\left(\mathbf{e}^{(z)}\right)^T \wedge \bigoplus_{\substack{\mathbf{e}^{(i)}=(Qu_i^T)^T \\ u_i \in \mathbf{U}(e)\backslash\{u_z\}}} P_{(z)}\left(\mathbf{e}^{(i)}\right)^T\right)\right\}.$$

Considering the second addend, a coarser upper bound to the argument of the $\max\{...\}$ operator can be derived observing that, given three binary vectors a, b, c, $wt\left(c \wedge (a \oplus b)\right) \leq wt\left(c \wedge (a \vee b)\right) = wt\left((c \wedge a) \vee (c \wedge b)\right)$, where $\vee$ denotes the binary OR. Thus, the second addend in the previous inequality can be replaced by the following quantity

$$\max_{\substack{z,i \in \mathbf{I}(e) \\ z \neq i}}\left\{wt\left(\bigvee_{\substack{(\mathbf{e}^{(i)})=(Qu_i^T)^T \\ u_i \in \mathbf{U}(e)\backslash\{u_z\}}}\left(P_{(z)}\left(\mathbf{e}^{(z)}\right)^T \wedge P_{(z)}\left(\mathbf{e}^{(i)}\right)^T\right)\right)\right\}$$

which can be further upper bounded (noting that $|\mathbf{U}(e) \backslash \{u_z\}| = t-1$) as

$$(t-1)\max_{\substack{z,i \in \mathbf{I}(e) \\ z \neq i}}\left\{wt\left(P_{(z)}\left(\mathbf{e}^{(z)}\right)^T \wedge P_{(z)}\left(\mathbf{e}^{(i)}\right)^T\right)\right\}.$$

Looking at the original equality set to compute the value of $\mathtt{upc}_z$, it holds that

$$\mathtt{upc}_z \geq \min_{z \in \mathbf{I}(e)} \left\{ wt \left(P_{(z)} \left(e^{(z)} \right)^T \right) \right\} +$$
$$- (t-1) \max_{\substack{z,i \in \mathbf{I}(e) \\ z \neq i}} \left\{ wt \left(P_{(z)} \left(e^{(z)} \right)^T \bigwedge P_{(z)} \left(e^{(i)} \right)^T \right) \right\}.$$

Therefore, it is easy to acknowledge that

$$\mathtt{min\_upc}_{flip} \geq \alpha - (t-1)\beta.$$

In the following we determine $\mathtt{max\_upc}_{no\,flip}$ as the highest value of an unsatisfied parity-check count $\mathtt{upc}_z$, $z \in \{0, \ldots, n-1\}$, when the value of the corresponding bit in the actual error vector is unset, i.e., $e_z = 0$.

In this case we aim at computing an upper bound for $\mathtt{upc}_z$, considering that $u_z \notin \mathbf{U}(e)$ (and thus $z \notin \mathbf{I}(e)$); we have

$$\mathtt{upc}_z = wt \left(\bigoplus_{\substack{e^{(i)} = (Qu_i^T)^T \\ u_i \in \mathbf{U}(e)}} P_{(z)} \left(e^{(i)} \right)^T \right) \leq t \max_{z,i \in \mathbf{I}(e)} \left\{ wt \left(P_{(z)} \left(e^{(i)} \right)^T \right) \right\} = t\gamma$$

We can therefore employ $t\gamma$ as an upper bound for the value of $\mathtt{max\_upc}_{no\,flip}$.

Recalling that the Q-decoder procedure will retrieve all the values of the error vector e in a single iteration if $\mathtt{max\_upc}_{no\,flip} < \mathtt{min\_upc}_{flip}$ and the bit-flipping threshold b is such that $b = \mathtt{min\_upc}_{flip}$, it is easy to acknowledge that the maximum number of errors tolerable by the code is constrained by the following inequality

$$t\gamma < \alpha - (t-1)\beta \Rightarrow t < \frac{\alpha + \beta}{\gamma + \beta}. \tag{2}$$

$\square$

A.2   Probabilistic Analysis of the First Iteration of the Q-Decoder

In the following, we model the number of differences between the guessed error vector $\hat{e}$, provided as output of the first iteration of the Q-decoder, and the actual error vector e, as a random variable T over the discrete domain of integers $\{0, \ldots, t\}$, $t \geq 0$ having a probability mass function $Pr\left[\mathsf{T} = \tau\right]$, $\tau = wt\left(\hat{e}^* \oplus e\right)$ depending on the decoding strategy and the LDPC code parameters.

To quantify the said probability, we consider the decoding procedure employed by the LEDAcrypt systems assuming that the hypothesis of Lemma 1 holds. Given the equivalence of the BF decoder and Q-decoder provided by this Lemma, for the sake of simplicity, we will reason on the application of one iteration of the BF decoder taking as input the $r \times n$ parity-check matrix $H' = HQ$ (assumed to be computed as a cancellation-free product between H and Q), the $1 \times n$ syndrome $s = (H' \hat{e}^T)^T$, and a null guessed error vector $\hat{e} = 0_{1 \times n}$. The code is assumed to be an LDPC code as in the LEDA cryptosystems, with $r = (n_0 - 1)p$, $n = n_0 p$, p a prime number, $n_0 \in \{2, 3, 4\}$, while m denotes the number of non-zero entries in each row/column of the $n \times n$ matrix Q, and $d_v n_0$ denotes the number of non-zero entries in each row/column of the $r \times n$ matrix H. As a consequence, each row/column of the parity-check matrix H' exhibits $d'_c = d_v n_0 m$ non-zero entries. This implies that each parity-check equation (i.e., row) of H' involves d'_c variables of the guessed error vector $\hat{e}$.

The quantification of the probability to observe a certain number of differences between the guessed error vector, provided as output of the first iteration of the decoder, and the actual error vector can be evaluated considering the number of correctly and wrongly flipped bits after the first iteration of the decoding algorithm. In turn, each of these numbers, can be quantified reasoning on the following joint probabilities: $p_{correct-unsatisfied}$ and $p_{incorrect-unsatisfied}$.

The joint probability $p_{correct-unsatisfied} = Pr\left[\hat{e}_j = e_j = 0; h_{ij} = 1, s_i = 1\right]$ can be stated as the likelihood of occurrence of the following events:

- $\{\hat{e}_j = e_j = 0\}$ refers to the event describing the j-th bit of the guessed error vector that does not need to be flipped;
- $\{h_{ij} = 1, s_i = 1\}$ refers to the event describing the i-th parity-check equation (i.e., H'_i) that is unsatisfied (i.e., $s_i = 1$) when the j-th variable in $\hat{e}$ (i.e., $\hat{e}_j$) is included in it.

It is easy to acknowledge that the above events occur if an odd number of the t asserted bits in the unknown error vector are involved in $d'_c - 1$ parity-check equations, thus

$$p_{correct-unsatisfied} = \sum_{j=1, j \, odd}^{\min[d'_c - 1, t]} \frac{\binom{d'_c - 1}{j} \binom{n - d'_c}{t - j}}{\binom{n-1}{t}}.$$

An analogous line of reasoning allows to quantify also the joint probability $p_{incorrect-unsatisfied} = Pr\left[\hat{e}_i \neq e_i; h_{ij} = 1, s_i = 1\right]$, which can be stated as the likelihood of occurrence of the following events:

- $\{\hat{e}_j \neq e_j\}$ refers to the event describing the j-th bit of the guessed error vector that needs to be flipped;
- $\{h_{ij} = 1, s_i = 1\}$ refers to the event describing the i-th parity-check equation (i.e., H'_i) that is unsatisfied (i.e., $s_i = 1$) when the j-th variable in $\hat{e}$ (i.e., $\hat{e}_j$) is included in it. So

$$p_{incorrect-unsatisfied} = \sum_{j=0, j \, even}^{\min[d'_c - 1, t-1]} \frac{\binom{d'_c - 1}{j} \binom{n - d'_c}{t - j - 1}}{\binom{n-1}{t-1}}.$$

The probability $\mathrm{p}_{\mathrm{correct}}$ that the upc based estimation deems rightfully a given $\hat{e}_j \neq e_j$ in need of flipping can be quantified as the probability that $\mathrm{upc} \geq b$, i.e.,

$$\mathrm{p}_{\mathrm{correct}} = \sum_{j=b}^{d'_v} \binom{d'_v}{j} \mathrm{p}_{\mathrm{incorrect-unsatisfied}}^{j} (1 - \mathrm{p}_{\mathrm{incorrect-unsatisfied}})^{d'_v - j} .$$

Analogously, we define the probability $\mathrm{p}_{\mathrm{induce}}$ as the probability that the upc based estimation deems a given $\hat{e}_j = e_j$ as (wrongly) in need of flipping as

$$\mathrm{p}_{\mathrm{induce}} = \sum_{j=b}^{d'_v} \binom{d'_v}{j} \mathrm{p}_{\mathrm{correct-unsatisfied}}^{j} (1 - \mathrm{p}_{\mathrm{correct-unsatisfied}})^{d'_v - j} .$$

Note that $\mathrm{p}_{\mathrm{correct}}$ is indeed the probability that the Q-decoder performs a correct flip at the first iteration, while $\mathrm{p}_{\mathrm{induce}}$ is the one of performing a wrong flip.

Thus, the probabilities of the Q-decoder performing $c \in \{0, \dots, t\}$ correct flips out of t or $w \in \{0, \dots, t\}$ wrong flips out of t can be quantified introducing the random variables f_{correct} and f_{wrong}, as follows

$$Pr\left[f_{\mathrm{correct}} = c\right] = \binom{t}{c} \mathrm{p}_{\mathrm{correct}}^{c} (1 - \mathrm{p}_{\mathrm{correct}})^{t-c} ,$$

$$Pr\left[f_{\mathrm{wrong}} = w\right] = \binom{n-t}{w} \mathrm{p}_{\mathrm{induce}}^{w} (1 - \mathrm{p}_{\mathrm{induce}})^{n-t-w} .$$

Assuming that the decisions on whether a given value $\hat{e}_j$ in $\hat{e}$ should be flipped or not are taken independently, i.e.,

$$Pr\left[f_{\mathrm{correct}} = c, f_{\mathrm{wrong}} = w\right] = Pr\left[f_{\mathrm{correct}} = c\right] \cdot Pr\left[f_{\mathrm{wrong}} = w\right],$$

we obtain the probability that the guessed error vector $\hat{e}$, at the end of the computation of the first iteration of the Q-decoder, differs from the actual error vector in $\tau \in \{0, \dots, t\}$ positions as follows

$$Pr\left[\mathrm{T} = \tau\right] = \sum_{i=t-\tau}^{t} Pr\left[f_{\mathrm{correct}} = i\right] \cdot Pr\left[f_{\mathrm{wrong}} = \tau + i - t\right].$$

This result permits us to estimate the probability of having a given number of errors $\tau \in \{0, \dots, t\}$ left to be corrected after the first iteration of the Q-decoder, since in this case the hypothesis on the independence of the decisions to flip or not to flip a given variable can be assumed safely.

References

1. LEDAtools (2019). https://github.com/LEDAcrypt/LEDAtools
2. Baldi, M., Barenghi, A., Chiaraluce, F., Pelosi, G., Santini, P.: LEDAkem: a post-quantum key encapsulation mechanism based on QC-LDPC codes. In: Lange, T., Steinwandt, R. (eds.) PQCrypto 2018. LNCS, vol. 10786, pp. 3–24. Springer, Cham (2018). https://doi.org/10.1007/978-3-319-79063-3_1

3. Baldi, M., Barenghi, A., Chiaraluce, F., Pelosi, G., Santini, P.: LEDAcrypt website (2019). https://www.ledacrypt.org/

4. Baldi, M., Bodrato, M., Chiaraluce, F.: A new analysis of the McEliece cryptosystem based on QC-LDPC codes. In: Ostrovsky, R., De Prisco, R., Visconti, I. (eds.) SCN 2008. LNCS, vol. 5229, pp. 246–262. Springer, Heidelberg (2008). https://doi.org/10.1007/978-3-540-85855-3_17

5. Becker, A., Joux, A., May, A., Meurer, A.: Decoding random binary linear codes in $2^{n/20}$: $1 + 1 = 0$ how improves information set decoding. In: Pointcheval, D., Johansson, T. (eds.) EUROCRYPT 2012. LNCS, vol. 7237, pp. 520–536. Springer, Heidelberg (2012). https://doi.org/10.1007/978-3-642-29011-4_31

6. Berlekamp, E., McEliece, R., van Tilborg, H.: On the inherent intractability of certain coding problems. IEEE Trans. Inf. Theory **24**(3), 384–386 (1978)

7. Bernstein, D.J.: Grover vs. McEliece. In: Sendrier, N. (ed.) PQCrypto 2010. LNCS, vol. 6061, pp. 73–80. Springer, Heidelberg (2010). https://doi.org/10.1007/978-3-642-12929-2_6

8. de Vries, S.: Achieving 128-bit security against quantum attacks in OpenVPN. Master's thesis, University of Twente, August 2016

9. Fabšič, T., Hromada, V., Stankovski, P., Zajac, P., Guo, Q., Johansson, T.: A reaction attack on the QC-LDPC McEliece cryptosystem. In: Lange, T., Takagi, T. (eds.) PQCrypto 2017. LNCS, vol. 10346, pp. 51–68. Springer, Cham (2017). https://doi.org/10.1007/978-3-319-59879-6_4

10. Fabšič, T., Hromada, V., Zajac, P.: A reaction attack on LEDApkc. IACR Cryptology ePrint Archive 2018:140 (2018)

11. Finiasz, M., Sendrier, N.: Security bounds for the design of code-based cryptosystems. In: Matsui, M. (ed.) ASIACRYPT 2009. LNCS, vol. 5912, pp. 88–105. Springer, Heidelberg (2009). https://doi.org/10.1007/978-3-642-10366-7_6

12. Gaborit, P.: Shorter keys for code based cryptography. In: Proceedings of International Workshop on Coding and Cryptography (WCC 2005), Bergen, Norway, pp. 81–90, March 2005

13. Gallager, R.G.: Low-Density Parity-Check Codes. MIT Press, Cambridge (1963)

14. Grassl, M., Langenberg, B., Roetteler, M., Steinwandt, R.: Applying Grover's algorithm to AES: quantum resource estimates. In: Takagi, T. (ed.) PQCrypto 2016. LNCS, vol. 9606, pp. 29–43. Springer, Cham (2016). https://doi.org/10.1007/978-3-319-29360-8_3

15. Grover, L.K.: A fast quantum mechanical algorithm for database search. In: Proceedings of 28th Annual ACM Symposium on the Theory of Computing, Philadephia, PA, pp. 212–219, May 1996

16. Guo, Q., Johansson, T., Stankovski, P.: A key recovery attack on MDPC with CCA security using decoding errors. In: Cheon, J.H., Takagi, T. (eds.) ASIACRYPT 2016. LNCS, vol. 10031, pp. 789–815. Springer, Heidelberg (2016). https://doi.org/10.1007/978-3-662-53887-6_29

17. Guo, Q., Johansson, T., Stankovski Wagner, P.: A key recovery reaction attack on QC-MDPC. IEEE Trans. Inf. Theory **65**(3), 1845–1861 (2019)

18. Hofheinz, D., Hövelmanns, K., Kiltz, E.: A modular analysis of the Fujisaki-Okamoto transformation. In: Kalai, Y., Reyzin, L. (eds.) TCC 2017. LNCS, vol. 10677, pp. 341–371. Springer, Cham (2017). https://doi.org/10.1007/978-3-319-70500-2_12

19. Jiang, H., Zhang, Z., Chen, L., Wang, H., Ma, Z.: IND-CCA-secure key encapsulation mechanism in the quantum random oracle model, revisited. In: Shacham, H., Boldyreva, A. (eds.) CRYPTO 2018. LNCS, vol. 10993, pp. 96–125. Springer, Cham (2018). https://doi.org/10.1007/978-3-319-96878-0_4

20. Jiang, H., Zhang, Z., Ma, Z.: Tighter security proofs for generic key encapsulation mechanism in the quantum random oracle model. Cryptology ePrint Archive, Report 2019/134, to appear in PQCrypto 2019 (2019). https://eprint.iacr.org/2019/134
21. Kachigar, G., Tillich, J.-P.: Quantum information set decoding algorithms. In: Lange, T., Takagi, T. (eds.) PQCrypto 2017. LNCS, vol. 10346, pp. 69–89. Springer, Cham (2017). https://doi.org/10.1007/978-3-319-59879-6_5
22. Karp, R.M.: Reducibility among Combinatorial Problems. In: Miller, R.E., Thatcher, J.W., Bohlinger, J.D. (eds.) Complexity of Computer Computations. The IBM Research Symposia Series. Springer, Boston, MA (1972)
23. Kobara, K., Imai, H.: Semantically secure McEliece public-key cryptosystems – conversions for McEliece PKC. In: Kim, K. (ed.) PKC 2001. LNCS, vol. 1992, pp. 19–35. Springer, Heidelberg (2001). https://doi.org/10.1007/3-540-44586-2_2
24. Lee, P.J., Brickell, E.F.: An observation on the security of McEliece's public-key cryptosystem. In: Barstow, D., Brauer, W., Brinch Hansen, P., Gries, D., Luckham, D., Moler, C., Pnueli, A., Seegmüller, G., Stoer, J., Wirth, N., Günther, C.G. (eds.) EUROCRYPT 1988. LNCS, vol. 330, pp. 275–280. Springer, Heidelberg (1988). https://doi.org/10.1007/3-540-45961-8_25
25. Leon, J.S.: A probabilistic algorithm for computing minimum weights of large error-correcting codes. IEEE Trans. Inf. Theory 34(5), 1354–1359 (1988)
26. May, A., Meurer, A., Thomae, E.: Decoding random linear codes in $\tilde{\mathcal{O}}(2^{0.054n})$. In: Lee, D.H., Wang, X. (eds.) ASIACRYPT 2011. LNCS, vol. 7073, pp. 107–124. Springer, Heidelberg (2011). https://doi.org/10.1007/978-3-642-25385-0_6
27. McEliece, R.J.: A public-key cryptosystem based on algebraic coding theory. DSN Progress Report, pp. 114–116 (1978)
28. Misoczki, R., Barreto, P.S.L.M.: Compact McEliece keys from Goppa codes. In: Jacobson, M.J., Rijmen, V., Safavi-Naini, R. (eds.) SAC 2009. LNCS, vol. 5867, pp. 376–392. Springer, Heidelberg (2009). https://doi.org/10.1007/978-3-642-05445-7_24
29. Misoczki, R., Tillich, J.P., Sendrier, N., Barreto, P.S.L.M.: MDPC-McEliece: new McEliece variants from moderate density parity-check codes. In: Proceedings of IEEE International Symposium on Information Theory (ISIT 2000), pp. 2069–2073, July 2013
30. Monico, C., Rosenthal, J., Shokrollahi, A.: Using low density parity check codes in the McEliece cryptosystem. In: Proceedings of IEEE International Symposium on Information Theory (ISIT 2000), Sorrento, Italy, p. 215, June 2000
31. National Institute of Standards and Technology: Post-quantum crypto project, December 2016. https://csrc.nist.gov/Projects/Post-Quantum-Cryptography
32. Niederreiter, H.: Knapsack-type cryptosystems and algebraic coding theory. Problems Control Inf. Theory 15, 159–166 (1986)
33. Nilsson, A., Johansson, T., Stankovski Wagner, P.: Error amplification in code-based cryptography. IACR Transactions on Cryptographic Hardware and Embedded Systems, 2019(1), pp. 238–258, November 2018
34. Prange, E.: The use of information sets in decoding cyclic codes. IRE Trans. Inf. Theory 8(5), 5–9 (1962)
35. Sendrier, N.: Decoding one out of many. In: Yang, B.-Y. (ed.) PQCrypto 2011. LNCS, vol. 7071, pp. 51–67. Springer, Heidelberg (2011). https://doi.org/10.1007/978-3-642-25405-5_4
36. Stern, J.: A method for finding codewords of small weight. In: Cohen, G., Wolfmann, J. (eds.) Coding Theory 1988. LNCS, vol. 388, pp. 106–113. Springer, Heidelberg (1989). https://doi.org/10.1007/BFb0019850

37. Tillich, J.: The decoding failure probability of MDPC codes. In: 2018 IEEE International Symposium on Information Theory, ISIT 2018, Vail, CO, USA, 17–22 June 2018, pp. 941–945 (2018)
38. Ueno, R., Morioka, S., Homma, N., Aoki, T.: A high throughput/gate AES hardware architecture by compressing encryption and decryption datapaths. In: Gierlichs, B., Poschmann, A.Y. (eds.) CHES 2016. LNCS, vol. 9813, pp. 538–558. Springer, Heidelberg (2016). https://doi.org/10.1007/978-3-662-53140-2_26

Introducing Arithmetic Failures to Accelerate QC-MDPC Code-Based Cryptography

Antonio Guimarães[1], Edson Borin[1], and Diego F. Aranha[1,2(✉)]

[1] Institute of Computing, University of Campinas, São Paulo, Brazil
antonio.junior@students.ic.unicamp.br, edson@ic.unicamp.br,
dfaranha@eng.au.dk
[2] Department of Engineering, Aarhus University, Aarhus, Denmark

Abstract. In this work, we optimize the performance of QC-MDPC code-based cryptosystems through the insertion of configurable failure rates in their arithmetic procedures. We present constant time algorithms with a configurable failure rate for multiplication and inversion over binary polynomials, the two most expensive subroutines used in QC-MDPC implementations. Using a failure rate negligible compared to the security level (2^{-128}), our multiplication is 2 times faster than NTL on sparse polynomials and 1.6 times faster than a naive constant-time sparse polynomial multiplication. Our inversion algorithm, based on Wu *et al.*, is 2 times faster than the original algorithm and 12 times faster than Itoh-Tsujii using the same modulus polynomial ($x^{32749} - 1$). By inserting these algorithms in a version of QcBits at the 128-bit quantum security level, we were able to achieve a speedup of 1.9 on the key generation and up to 1.4 on the decryption time. Comparing with variant 2 of the BIKE suite, which also implements the Niederreiter Cryptosystem using QC-MDPC codes, our final version of QcBits performs the uniform decryption 2.7 times faster.

Keywords: Code-based cryptography · QC-MDPC ·
Polynomial inversion · Polynomial multiplication · Failure insertion

1 Introduction

The emergence of quantum computers is pushing an unprecedented transition in the field of public-key cryptography. Classical cryptosystems are based on the hardness of solving integer factorization and discrete logarithms, problems that can be solved in polynomial time using quantum computers [32]. To replace them, several algorithms in the field of post-quantum cryptography are being considered [28]. Among them, code-based cryptography is a promising candidate, primarily represented by the McEliece cryptosystem [24] and its variations, with security based on the hardness of decoding linear codes [3]. There are many

M. Baldi et al. (Eds.): CBC 2019, LNCS 11666, pp. 44–68, 2019.
https://doi.org/10.1007/978-3-030-25922-8_3

proposals for families of codes to instantiate the McEliece cryptosystem, but two excel in terms of security and performance: Goppa [15] and QC-MDPC [25].

Goppa codes were originally used by Robert McEliece in his cryptosystem [24] with great decoding performance, but very large keys. Most of the attempts of reducing the key size of Goppa codes resulted in structural vulnerabilities [12]. QC-MDPC codes were introduced in 2013 by Misoczki et al. [25] as one of these attempts, created as a variant of QC-LDPC codes to achieve very compact keys while avoiding the structural vulnerabilities that afflicted LDPC codes. Two major drawbacks, however, are the deteriorated performance compared to Goppa codes and a non-negligible decoding failure rate (DFR).

One of the main features of QC-MDPC cryptosystems is the use of the Quasi-Cyclic (QC) structure, which enables cryptosystems to be entirely implemented using arithmetic over binary polynomials. The keys, messages, and errors are represented as binary polynomials and three arithmetic operations are executed over them: addition, multiplication, and inversion. The first is efficient, but the others usually represent over 90% of the execution time of the cryptosystem. Algorithms for arithmetic over binary polynomials are very well-known and studied. Hence, new generic optimizations for them are usually restrained to the implementation aspect only.

In this work, we take a different approach to optimize binary field arithmetic, focusing specifically on the case of QC-MDPC code-based cryptography. We first selected constant-time algorithms from the literature that better exploit the special characteristics of QC-MDPC polynomials on modern computer architectures, such as the large size and relatively low density. Then, we modified the algorithms to accept configurable parameters that greatly accelerate them at the cost of introducing a negligible probability of failure depending on the input. Finally, we defined methods to correlate this probability of failure (or failure rate) of each algorithm with the impact on performance. In this way, we present the following contributions.

- We introduce the concept of using arithmetic subroutines with a controlled failure rate to accelerate implementations of QC-MDPC code-based cryptosystems.
- We present constant-time algorithms for multiplication and inversion over binary polynomials that operate with configurable failure rates.
- We define methods to obtain a correlation between failure rate and performance improvement for each algorithm.
- We show that these algorithms provide a significant performance improvement while introducing an arithmetic failure rate that is negligible compared to the security level of the cryptosystem.

This paper is organized as follows: Sect. 2 presents the basic concepts of code-based cryptography and QC-MDPC codes; Sect. 3 discusses the techniques used to accelerate the multiplication and inversion algorithms; Sect. 4 shows the performance results; Sect. 5 presents the related work; and Sect. 6 finalizes the paper with our conclusions.

2   Code-Based Cryptography

The field of code-based cryptography started with the publication of the McEliece Cryptosystem [24] using Goppa Codes in 1978. The cryptosystem presented great performance at the time (being faster than RSA [30]) and a solid security basis [3]. A major problem, which impaired the acceptance of the cryptosystem, was the size of the keys. Many were the attempts to reduce its key size. Both Goppa derivatives and other families of codes were proposed, but most of them resulted in structural vulnerabilities [12]. In 2000, Monico *et al.* [26] suggested the use of Low-Density Parity-Check (LDPC) codes [14] in the McEliece cryptosystem. In 2002, the idea of using a parity check matrix with a Quasi-Cyclic (QC) structure for LDPC codes [22] brought great advantages in terms of performance and key size. The quasi-cyclic structure is exemplified below. For each circulating block (the example has 2), the i-th row is the $(i-1)$-th row rotated one bit to the right. Originally, all the keys of the cryptosystem were matrices. This structure allowed them to be represented by its first row only and to be treated as polynomials over $x^r - 1$, where r is the size of each circulating block.

$$\begin{pmatrix} 1\,0\,0\,1\,0\ \ 0\,0\,0\,1\,1 \\ 0\,1\,0\,0\,1\ \ 1\,0\,0\,0\,1 \\ 1\,0\,1\,0\,0\ \ 1\,1\,0\,0\,0 \\ 0\,1\,0\,1\,0\ \ 0\,1\,1\,0\,0 \\ 0\,0\,1\,0\,1\ \ 0\,0\,1\,1\,0 \end{pmatrix}$$

The proposal of using a quasi-cyclic structure did not intend to solve the structural vulnerability problems of LDPC codes, but it created the basis for the development of new families of codes that would avoid them. In 2013, Misoczki *et al.* [25] proposed the use of Quasi-Cyclic Moderate Density Parity Check (QC-MDPC) codes, a derivative of QC-LDPC using higher density parity check matrices. The cryptosystem kept the compact size of the keys from the LDPCs cryptosystems and avoided structural vulnerabilities by increasing the density of the parity matrix. The performance, however, was significantly worse than Goppa codes, especially in the key generation and decryption procedures.

In 2016, Chou published QcBits [8], the first constant-time implementation of QC-MDPC codes, and the fastest at the time. Constant-time execution is an important feature since side-channel attacks against implementations of code-based cryptography have been frequently explored in the literature [11,31,36]. In 2017, Aragon *et al.* published the BIKE suite [1] containing 3 key encapsulation schemes using QC-MDPC codes. BIKE is the main representative of QC-MDPC codes in NIST's standardization project [28] and its Variation 2 implements the same cryptosystem as QcBits, the Niederreiter Cryptosystem [27].

2.1   Implementation of a QC-MDPC Code-Based Cryptosystem

Algorithms 1, 2 and 3 describe the processes for key generation, encryption and decryption, respectively. The cryptosystem implemented is the Niederreiter Cryptosystem [27], a simpler variation of McEliece. The algorithms employ the polynomial notation only since our proposals are specifically focused on the arithmetic over binary polynomials. The parameters R, W and T are defined by the target security level. R is the degree of the modulus polynomial, W is the Hamming weight of the key, and T is the Hamming weight of the error polynomials. Their values at the 128-bit quantum security level are 32749, 274 and 264, respectively [1]. The function $GeneratePolynomial$ generates a binary polynomial with the specified Hamming weight and maximum degree. The encryption process is structured to be part of a Key Encapsulation Mechanism (KEM), but it could also be modified to encrypt a message. The error polynomials e_0 and e_1 are randomly generated and can be used as a key for a symmetric cryptosystem.

Algorithm 1: Key Generation.

   Input   : $GeneratePolynomial$, R and W

   Output: $PrivateKey$ and $PublicKey$

1 $H_0 \leftarrow GeneratePolynomial(MaxDegree = R - 1, HammingWeight = \frac{W}{2})$

2 $H_1 \leftarrow GeneratePolynomial(MaxDegree = R - 1, HammingWeight = \frac{W}{2})$

3 $PublicKey \leftarrow H_0^{-1} \times H_1$

4 $PrivateKey \leftarrow (H_0, H_1)$

Algorithm 2: Encryption.

   Input   : $G \leftarrow PublicKey$, R, T and $GeneratePolynomial$

   Output: $Ciphertext$ and Key

1 $e_0 \leftarrow GeneratePolynomial(MaxDegree = R - 1, HammingWeight = \frac{T}{2})$

2 $e_1 \leftarrow GeneratePolynomial(MaxDegree = R - 1, HammingWeight = \frac{T}{2})$

3 $Ciphertext \leftarrow e_1 \times G + e_0$

4 $Key \leftarrow (e_0, e_1)$

The decryption process is composed of syndrome calculation (lines 3 and 13) and the decoding algorithm (lines 5 to 12), applied iteratively. There are many published algorithms for the decoding part, in this case we are using a simple version of Gallager's bit-flipping decoding [14]. The function $TransposePoly$ obtains the polynomial representing the column for a Quasi-Cyclic matrix from the polynomial representing the row. The polynomial sum is an integer polynomial and the function $IntegerPolynomialAddition$ interprets w as an integer polynomial and adds it to sum. The function $CalculateThreshold$ computes the threshold used to define which bits probably belong to the error polynomials and the method used to determine it varies with the implementation.

Algorithm 3: Decryption using the bit-flipping algorithm.

Input : H and c
Output: e_0 and e_1

1 $e_0 \leftarrow 0, \; e_1 \leftarrow 0$
2 $H_0'(x) \leftarrow TransposePoly(H_0(x)), \; H_1'(x) \leftarrow TransposePoly(H_1(x))$
3 $s \leftarrow (H_0 \times c)$
4 **while** $s \neq 0$ **do**
5 $\quad$ **for** $j = 0 \rightarrow 1$ **do**
6 $\quad\quad$ $sum \leftarrow 0$
7 $\quad\quad$ **foreach** $monomial \; x^i \in H_j'(x)$ **do**
8 $\quad\quad\quad$ $w \leftarrow s \times x^i$
9 $\quad\quad\quad$ $sum \leftarrow IntegerPolynomialAddition(sum, w)$
10 $\quad\quad$ $Threshold \leftarrow CalculateThreshold(s)$
11 $\quad\quad$ **foreach** $monomial \; x^i \in sum(x)$ **do**
12 $\quad\quad\quad$ **if** $[x^i](sum(x)) > Threshold$ **then** $e_j \leftarrow e_j + x^i$;
13 $\quad$ $s \leftarrow (H_0 \times (e_0 + c)) + (H_1 \times e_1)$

Since fully understanding these algorithms is not essential here, we are more interested in showing how binary field arithmetic works in each phase of the cryptosystem. Key generation uses two multiplications and an inversion over binary polynomials. Encryption uses a multiplication and an addition. The decryption algorithm explicitly uses two multiplications and two additions (line 13) per iteration, but the monomial multiplications in line 8 and the monomial additions in line 12 can also be implemented as just one polynomial multiplication and addition, respectively. In this way, the decryption takes four multiplications and four additions per iteration, plus one multiplication at the beginning (line 3).

3   Algorithms with a Configurable Failure Rate

As we showed in Sect. 2.1, the three main arithmetic operations used in QC-MDPC implementations are addition, multiplication, and inversion over binary polynomials. The addition is implemented as a sequence of XOR instructions in most architectures, very efficient operations which leave very little space for software optimization, even considering the introduction of failure. Thus, our work focus on the multiplication and inversion operations. In this section, we present algorithms for these operations accepting a configurable failure rate.

We benchmarked our implementations on an Intel i7-7820X processor with Hyper-Threading and TurboBoost disabled to improve the reproducibility of the experiments [4]. We implemented the algorithms in C language using *intrinsics* for the AVX-512 instruction set extension and compiled with GCC 7.3.1. We used inverted binary and sparse representations of polynomials (examples in Appendix B) with a maximum degree of 32,748 and Hamming weight of 137. These values are defined based on the parameters for 128-bit quantum security level in QC-MDPC implementations.

We implemented the conditional statements of the algorithms in constant-time using conditional operations. Examples of vectorized constant-time implementations for If statements are provided in Appendix A. We choose to implement our code using AVX-512 instructions to allow a direct comparison with other highly optimized implementations of QC-MDPC cryptosystems, such as the *Additional Implementation* in BIKE. Moreover, our implementation benefits from the possibility of implementing constant-time operations using mask registers, as also shown in Appendix A. This feature was only introduced recently in Intel architectures, but it is common in others (e.g. ARM A32).

3.1 Polynomial Inversion

We based our inversion method on the inversion algorithm by Wu *et al.* [37]. Algorithm 4 shows its original form. It was created as an extended version of Stein's algorithm [35], which avoids the extra costs of calculating degrees of polynomials that is common in Extended Euclidean Algorithms [19]. It is similar to the algorithm due to Brunner *et al.* [6], differing by the testing of the least significant bit instead of the most significant one. While theoretically equivalent, this modification makes the implementation of Wu *et al.* much simpler than Brunner *et al.*'. Moreover, the algorithm was also designed to be regular for hardware implementation, which makes it easier to implement in constant-time. The algorithm receives the binary polynomials r, s and u and calculates $\frac{u}{r}$ $(mod\ s)$. Since we are interested in the inversion, we use $u = 1$, while r is the target polynomial and s is the modulus polynomial. The algorithm also receives the size of the polynomials, N.

Algorithm 4: Wu *et al.* Inversion Algorithm [37].

Input : r, s, u and N
Output: $v = \frac{u}{r}$ $(mod\ s)$
1 $v \leftarrow 0,\ \delta \leftarrow -1,\ g \leftarrow s$
2 **for** $i = 0$ **to** $2 \times N$ **do**
3   **if** $r_0 = 1$ **then**
4   **if** $\delta < 0$ **then**
5           $(r, s, u, v) \leftarrow (r + s, r, u + v, u)$
6           $\delta \leftarrow -\delta$
7   **else**
8           $(r, u) \leftarrow (r + s, u + v)$
9     $(r, u) \leftarrow (r/x, (u/x)_g)$
10    $\delta \leftarrow \delta - 1$

The algorithm iterates over the polynomial r, verifying the existence of a 0-degree monomial and dividing it per x until the monomial is found. Once it occurs, the if (line 3) is executed and the resultant polynomial v is modified. As previously mentioned, compared to inversion approaches based on Euclidean

Algorithm, one of the main advantages of Stein's algorithm and its derivatives is avoiding expensive degree comparisons between polynomials. To enable the possibility of introducing failures in the algorithm, we had to take a step back on this advantage. We reintroduced a degree verification in the algorithm as a search for the lowest degree monomial of r. This way, we produced the modified version of Wu *et al.* in Algorithm 5. In Appendix B, we discuss how we implemented the degree verification efficiently.

Algorithm 5: Modified version of the Wu *et al.* Inversion Algorithm [37].

Input : r, s, u, N and F
Output: $v = \frac{u}{r} \ (mod \ s)$

1   $v \leftarrow 0, \delta \leftarrow -1, g \leftarrow s$
2   **for** $i = 0$ **to** $F \times 2N$ **do**
3      $b \leftarrow SmallestMonomialDegree(r)$
4      $(r, u) \leftarrow (r/x^b, (u/x^b)_g)$
5      $\delta \leftarrow \delta - b$
6   **if** $r \neq 0$ **then**
7   **if** $\delta < 0$ **then**
8              $(r, s, u, v) \leftarrow (r + s, r, u + v, u)$
9              $\delta \leftarrow -\delta$
10   **else**
11              $(r, u) \leftarrow (r + s, u + v)$

The condition (now in line 6) remained unchanged, a function to find the monomial with the smallest degree was inserted (line 3), the divisors in line 4 were changed from x to x^b, and, finally, the number of iterations is now reduced by a factor $0 < F \leq 1$. Using $F = 1$ would handle all inputs correctly, but without performance gains. Using lower values of F accelerates the algorithm, but it also inserts the possibility of failure. In this way, our modification exposed a parameter F that allows us to control the trade-off between performance and failure rate in the algorithm. We need now to define a method to precisely associate these two measures and to obtain a good value of F for our case.

Since F changes the number of loop iterations directly, the performance speedup is simply $1/F$. Defining the Failure Rate (FR) in terms of F is significantly more difficult. First, using $r_{0,i}$ as the value of r_0 in line 3 at the i-th iteration of Algorithm 4, we define a polynomial $R(x) = \sum_{i=0}^{2 \times N} r_{0,i} \, x^i$. The failure rate can be defined as the probability (function P) of $R(x)$ having a Hamming Weight (HW) greater than $(F \times 2 \times N)$, i.e $FR = P(HW(R(x)) > (F \times 2N)$. To solve this correlation exactly, we would have to consider the number of possible $R(x)$ polynomials and the probability of each one being generated by the algorithm. A simpler approach that results in a good approximation is to consider the probability of each monomial value individually.

Supposing that each monomial coefficient of R has an independent probability p of being 1, we can use the Binomial Expansion [18] to obtain the approximation in Eq. 1. Further supposing that $p = 0.5$ and using $N = 32749$, we obtain the chart of Fig. 1, which correlates the Failure Rate with the parameter F and give us the value $F = 0.5254$ to achieve a negligible failure rate. Using this value of F, the cost of Wu *et al.* reduces from 39,747,301 to 20,773,925 cycles, which represents a speedup of 1.9.

$$FR = P(HW(R(x)) > (F \times 2N)) \approx 1 - \sum_{i=0}^{\lceil F \times 2N \rceil} \binom{2N}{i}(1-p)^{(2N-i)}p^i \quad (1)$$

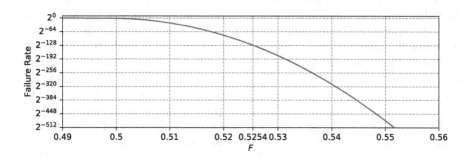

Fig. 1. Correlation between failure rate and parameter F in the inversion algorithm.

In this way, we defined the method to correlate Failure Rate and Performance Speedup using the parameter F and found a value that fits our case. However, we did it based on the assumption that each monomial coefficient of $R(x)$ has an independent probability $p = 0.5$ of being 1. To show that this is a good estimation for any input polynomials, we define the recurrence relation in Eq. 2 for the probability of $r_{0,i} = 1$. Analyzing it, we have that the probability in each iteration is either preserved or modified by a XOR operation. Thus, we have that if $P(r_{j,0} = 1) < 0.5$ then $P(r_{j,i} = 1) \leq 0.5$ for all $0 \leq i \leq 2N$, which satisfies our assumption. This result is proved in Appendix C. The approximation is not as tight as it could for low-density polynomials, but the resulting F parameter is sufficiently close to the optimization limit (0.5).

$$P(r_{j,i} = 1) = \begin{cases} P(r_{j,0} = 1), & \text{if } i = 0 \\ P(r_{0,i-1} = 1) \times P(s_{j+1,i-1} \oplus r_{j+1,i-1} = 1) & \\ \quad + P(r_{0,i-1} = 0) \times P(r_{j+1,i-1} = 1), & \text{if } i > 0 \end{cases} \quad (2)$$

Experimenting with Higher Failure Rates. Another optimization through introducing the possibility of failure in our inversion algorithm concerns the operations over the polynomials r and s. When a polynomial is large enough to need more than one word to be stored, which is our case, any operation over the polynomial is implemented iteratively over the array of words that stores it. In the

Wu *et al.* algorithm, the polynomial r always converges to zero, which implies in a degree reduction of the polynomial along the iterations of the algorithm. As the higher part of the binary representation of the polynomial becomes zeros, it does not need to be processed anymore. In constant-time implementations, however, all the words belonging to the array are always processed. In this way, we cannot check whether the higher parts are zeros or not and only process it based on this information. Nevertheless, we can estimate the degree reduction of the polynomials to decide if the higher parts need to be processed. This estimation aims to cover only the majority of the cases, but not all of them. Therefore, a failure rate can also be explored at this point.

For this modification, however, we did not define a strict correlation between the failure rate and the performance level. Instead, we estimate the failure rate through experimental data targeting at a 10^{-8} failure rate. Since the cryptosystems already have a global failure rate (the DFR) of around 10^{-7} in modern implementations, we can also introduce algorithms for the arithmetic operations that fail with a small but non-negligible probability. If we guarantee that this probability is at least one order of magnitude smaller than the DFR, then the impact on the global failure rate of the cryptosystem will be almost negligible.

To achieve a 10^{-8} failure rate, we measured the degree of 10^8 polynomials along the iterations of the algorithm. Figure 2 shows the minimum, maximum and average degrees of the polynomials. The step-function curve represents the upper bound estimation applied to achieve a failure rate smaller than 10^{-8}. It is a step-function because we only choose to process or not entire words of 512 bits. The zoomed portion shows that there is no intersection between the upper bound and maximum value curves.

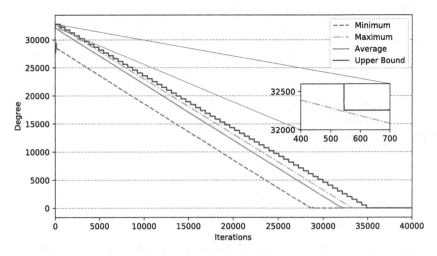

Fig. 2. Minimum, maximum and average degrees of r for 10^8 randomly generated polynomials along the iterations of Algorithm 11. The step-function curve represents an estimated upper bound to achieve a 10^{-8} failure rate.

A similar optimization can be applied to the polynomials u and v. The convergence of v to the inverse of r is similar to the convergence of r to zero. While r has its degree reduced along the iterations, v grows from the higher part through the insertions of 0-degree monomials followed by modular divisions. In this way, the lower part of v (and, consequently, u) is composed of zeros in the first iterations and, hence, does not need to be processed. We also estimated the degree of the lowest monomial of u along the iterations and established a lower bound to achieve a 10^{-8} failure rate. Figure 3 shows the results. An exception in the values of the lower part is the 0-degree monomial, which is not zero from the beginning. We eliminated it by dividing u per x before the algorithm, and multiplying the result by x afterward.

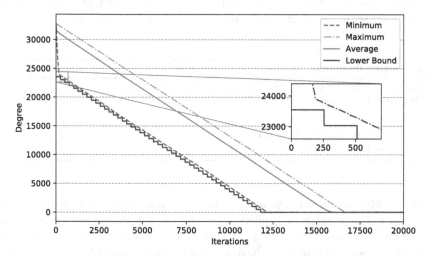

Fig. 3. Minimum, maximum and average degrees of u for 10^8 randomly generated polynomials along the iterations of Algorithm 11. The step-function curve represents an estimated lower bound to achieve a 10^{-8} failure rate.

The result of the upper and lower bound estimations for the degree of, respectively, r and u, was that r has its degree reduced by at least 0.94 per iteration starting from 32,768; while u has the degree of its smallest monomial decreased by at most 2 per iteration starting from 23,552. Using these estimations for the upper and lower bounds, we further reduced the number of cycles needed to calculate the inverse in our implementation from 20,773,925 to 14,979,764, which represents an overall speedup of 2.65.

Once we inserted the possibility of non-negligible failures, we also need to provide the methods to detect it. Ideally, we would detect the failure before trying to calculate the inverse, which would avoid a useless execution of this expensive procedure. Unfortunately, it is not simple to describe a correlation between the input and the occurrence of a failure. However, after the execution of the algorithm, it is easy to detect whether the inverse is correct or not by verifying if the product between the result and the input is the identity (one). This verification is already done by most implementations of code-based cryptography and is shown in line 4 of Algorithm 1.

3.2 Polynomial Multiplication

We start with a very basic algorithm that multiplies the first operand by each monomial taken from the second one and adds the results. Algorithm 6 shows this procedure. For generic polynomials, this algorithm is usually outperformed by specialized instructions, such as the carryless multiplier introduced by Intel (`PCLMULQDQ`). However, the algorithm has good performance if at least one of the polynomials has a relatively low density. As a general rule, this algorithm will outperform the specialized instructions if the bit density of at least one of the polynomials is less than the inverse of the processor word size.

Algorithm 6: Sparse multiplication algorithm.

Input   : A sparse polynomial P_1 and a polynomial P_2
Output: The result polynomial P_R

1   $P_R \leftarrow 0$
2   **foreach** *monomial* $m_i \in P_1$ **do**
3 | $P_R \leftarrow P_R + P_2 \times m_i$

Since additions are very cheap, the cost of the procedure can be estimated as the number of monomials multiplied by the cost of each *monomial* × *polynomial* multiplication. Considering that it is not possible to reduce the number of monomials, we focused on optimizing the monomial multiplication. Our modulus polynomial has the format $(x^n - 1)$, with $n \geq 1$, thus, the monomial multiplication can be implemented as just a rotation of the binary representation of the polynomial. This rotation is very simple to implement when the polynomial is small enough to be stored inside just one word of the architecture. For larger polynomials, however, it is necessary to perform a rotation of the words followed by a rotation with carry inside the words.

Algorithm 7 shows how the operation can be implemented in constant-time. It is based on the implementation used by QcBits [8]. M is the degree of the monomial, P is the polynomial in the binary representation, *wordSize* is the word size in which a shift or rotation operation can be executed in the architecture, and M_{max} is an upper bound for M. This procedure has complexity $\mathcal{O}(log_2(M_{max}))$. It should be noted that M in Algorithm 7 is the degree itself. In Algorithms 6, 8, 9 and 10, however, m_i is the entire monomial $m_i = c_i x^{M_i}$, where ($c_i \in 0, 1$) is the coefficient and M_i is the degree.

There are two ways of improving the performance of this algorithm: to increase the word size, which would rely on hardware modifications; or to decrease the upper bound of M, which we explore in this work. In the original version, the algorithm multiplies the polynomial with higher density P_2 by

---

Algorithm 7: Constant-time implementation of a *Monomial* × *Polynomial* multiplication

---

Input : M, P, $wordSize$, M_{max}
Output: $P_{out} = x^M \times P$

1   $P_{out} \leftarrow P$
    /* Word rotations   */
2   **for** $i \leftarrow \lfloor log_2(M_{max}) \rfloor$ **to** $log_2(word\_size) + 1$ **by** -1 **do**
3   **if** $M \wedge 2^i$ **then**
4         $P_{out} \leftarrow P_{out} \lll 2^i$ // Implemented through word copies
     /* Rotation inside the words (with carry)   */
5   $P_{out} \leftarrow P_{out} \lll (M \wedge (wordSize - 1))$

---

the absolute values of each monomial belonging to the sparse polynomial P_1. Our first modification, presented in Algorithm 8, is to store the result of the previous multiplication and use it to reduce the monomial exponent of the following multiplications. This is basically an application of Horner's rule. For example: if we want to multiply P_{ex} by $(x^{13} + x^7 + x^2)$, Algorithm 6 would calculate $(P_{ex} \times x^{13}) + (P_{ex} \times x^7) + (P_{ex} \times x^2)$ while Algorithm 8 computes $(P_{ex} \times x^2) \times ((x^5 \times (x^6 + 1)) + 1)$. The number of operations is the same, but Algorithm 8 reduces the value of M_{max} from 13 to 6. This reduction is more significant the higher the number of monomials and, in a non-constant-time implementation, it might provide an immediate gain with this modification.

Algorithm 8: Sparse multiplication algorithm using the relative degree of the monomials.

Input : A sparse polynomial P_1 and a polynomial P_2
Output: The result polynomial P_R

1   $P_R \leftarrow 0$
2   **foreach** *monomial* $m_i \in P_1$ **do**
3   **if** $i = 0$ **then**
4         $P_2 \leftarrow P_2 \times m_0$
5   **else**
6         $P_2 \leftarrow P_2 \times \frac{m_i}{m_{i-1}}$
7      $P_R \leftarrow P_R + P_2$

On constant-time implementations, if we consider the worst-case scenario, the upper bound M_{max} is close to the maximum degree of the monomials, resulting in almost no performance improvement. At this point, we introduce a trade-off between performance and failure rate by defining an upper bound M_{max} which does not cover all the cases. Using 512-bit words and 32749-bit polynomials, the possible values for M_{max} are 2^i for $9 \leq i \leq 15$. While this already represents a controlled failure rate, its granularity and the correlation with performance are not good enough. Thus, we produce a third version of the algorithm, depicted in Algorithm 9. If we set, for example, $M_{max} = 512$, we would have a higher performance improvement, but our failure rate in Algorithm 8 would be almost 100%. The algorithm fails because at least one pair of consecutive monomials has a difference between their exponents greater than the defined upper bound. In Algorithm 9, we introduce an auxiliary polynomial, P_A, which is specially constructed by inserting one or more monomials between those pairs, until the upper bound is respected and the failure is, consequently, eliminated. When adding the results to P_R (line 9 of Algorithm 9), the intermediate results generated by monomials belonging to P_A are ignored, not affecting the final product. In this way, P_A enables the algorithm to perform the multiplications in which it would fail. Algorithm 10 shows how to construct the auxiliary polynomial P_A in constant time.

Algorithm 9: Sparse multiplication algorithm using the relative degree of the monomials and an auxiliary polynomial.

Input : A sparse polynomial P_1, a polynomial P_2, M_{max} and HW_{P_A}
Output: The result polynomial P_R

1   $P_R \leftarrow 0$
2   $P_A \leftarrow ConstructPA(P_1, M_{max}, HW_{P_A})$
3   **foreach** *monomial* $m_i \in (P_1 + P_A)$ **do**
4   **if** $i = 0$ **then**
5         $P_2 \leftarrow P_2 \times m_0$
6   **else**
7         $P_2 \leftarrow P_2 \times \frac{m_i}{m_{i-1}}$
8   **if** $m_i \notin P_A$ **then**
9         $P_R \leftarrow P_R + P_2$

The number of monomials in P_A also needs to be fixed to preserve the constant-time execution. In this way, we have to define the Hamming weight of P_A, HW_{P_A}, for a given upper bound M_{max}, to achieve a desired failure rate. To define this failure rate in terms of HW_{P_A} and M_{max}, we counted the number of polynomials that would either fail or succeed under each of the possible parameters. To do that, the problem was formulated as the problem of counting the number of restricted compositions of an integer. A k-composition of n is an ordered sum of k positive integers that results in n [34]. A composition is an ordered partition, for example: $(2 + 5 + 3)$ and $(2 + 3 + 5)$ are the same partition of 10, but they are two different 3-compositions of 10.

The number of possible polynomials for each key of the cryptosystem is equal to the number of $W/2$-compositions of R, where W and R are the security parameters. To find the failure rate for the multiplication algorithm, we need a restriction to the composition count. For each part p of a composition, we define the number of *M-violations* as $\lfloor \frac{p}{M} \rfloor$. The total number H of *M-violations* of a

---

Algorithm 10: Construction of the auxiliary polynomial P_A

    Input : P_1, M_{max} and HW_{P_A}
    Output: P_A

1   $P_A \leftarrow 0$
2   **for** $i \leftarrow 0$ **to** HW_{P_A} **do**
3   **foreach** *monomial* $m_i \in (P_1 + P_A)$ **do**
4   **if** $i = 0$ **then**
5   **if** $m_i \geq x^{M_{max}}$ **then**
6               $P_A \leftarrow P_A + x^{M_{max}-1}$
7   **break**
8   **else**
9   **if** $\frac{m_i}{m_{i-1}} \geq x^{M_{max}}$ **then**
10               $P_A \leftarrow P_A + m_{i-1} \times x^{M_{max}-1}$
11   **break**
12   **if** $HammingWeight(P_A) \leq i$ **then**
          /* Adds monomials with degrees greater than degree of P_1 */
13         $P_A \leftarrow P_A + x^{Degree(P_1)+i}$

---

composition is the sum of the number of *M-violations* of each part. Defining a k-composition of n with at most H *M-violations* as an (M, H)-restricted k-composition of n, the failure rate for the multiplication algorithm is

$$\left(1 - \frac{\text{number of } (M_{max}, HW_{P_A}) - \text{restricted } \frac{W}{2} - \text{compositions of } R}{\text{number of } \frac{W}{2} - \text{compositions of } R}\right)$$

To compute the number of compositions, we define the recurrence relation in Eq. 3, where $C[k, n, H]$ is the number of k-compositions of n with exactly H *M-violations*. Using dynamic programming, we solve the recurrence and produced the chart in Fig. 4. We did not analyze values of M_{max} greater than 4096 (2^{12}) because their performance considering $HW_{P_A} = 0$ was already worse than $M_{max} = 512$ with $HW_{P_A} = \lfloor \frac{R}{M_{max}} \rfloor - 1$, which is a case without failure.

$$C[k, n, H] = \begin{cases} 1, & \text{if } k = 1 \text{ and } H = \lfloor \frac{n}{M} \rfloor \\ \sum_{i=max(k-1,n+1-M(H+1))}^{n-1} C[k-1, i, H - \lfloor \frac{n-i}{M} \rfloor], & \text{if } k > 1 \\ 0, & \text{otherwise} \end{cases} \quad (3)$$

We estimated the performance cost of each M_{max} presented in the chart using the HW_{P_A} necessary to obtain negligible failure rates. Table 1 shows the performance results. The best estimated result was for $M_{max} = 1024$, closely followed by $M_{max} = 512$. Despite being slightly slower, we recommend the use of $M_{max} = 512$ with $HW_{P_A} = 47$ because it facilitates the implementation. Over the initial multiplication algorithm, these parameters provide a speedup of 1.63, reducing the execution time from 130,023 cycles to 76,805 cycles.

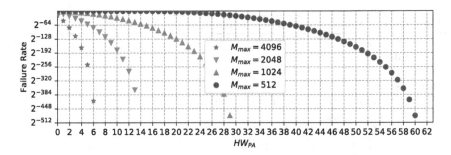

Fig. 4. Correlation between failure rate and the value of the parameters M_{max} and HW_{P_A} in the multiplication algorithm.

Table 1. Estimated and measured execution cost (cycles) of the multiplication according to the value of M_{max}.

M_{max}	Multiplication cost	HW_{P_A}	Estimated total cost	Total cost
512	56,388	47	75,733	76,805
1024	65,747	20	75,345	76,094
2048	74,645	9	79,549	-
4096	87,014	4	89,555	-

Sorting the Degree of Monomials. Our multiplication algorithm requires the monomials of the sparse polynomial to be sorted by their degree. Thus, a constant-time sorting algorithm is necessary. We implemented a vectorized constant-time version of Counting Sort that sorts by degree and creates the polynomial P_A at the same time. Its execution takes 66,833 cycles, which raises the total cost of the multiplication to 130,464 cycles if we consider that only one multiplication will be executed over a certain sparse polynomial. That is the case in the encryption of QC-MDPC cryptosystems, and, in this step, we have no performance gain on the multiplication. On the decryption, however, the same sparse monomial is used in many multiplications. In the constant-time case in QcBits, for example, each key needs to be sorted once and is then used in 22 multiplications. In this way, the sorting cost per multiplication is about 3,000 cycles, raising the multiplication cost to only 79,843. Moreover, the sorting can be pre-processed at the key-generation procedure. Alternatives to overcome this problem are to design a polynomial generation algorithm that provides monomials already sorted by the degree or to use more efficient constant-time sorting algorithms, such as Bernstein's djbsort [5].

Consideration About the New `VPCLMULQDQ` *Instruction.* An AVX-512 version of Intel's `PCLMULQDQ` instruction was announced as part of the Ice Lake architecture. Drucker *et al.* [9] estimated that this instruction will bring a speedup of about 2 times to multiplication algorithms that employ it, which is the case of the NTL Library, for example. This forthcoming improvement, however, does not reduce the importance of our multiplication algorithm. Our algorithm not only accelerates the generic sparse multiplications but also the Unsatisfied Parity-Check Counting (lines 7 and 8 of Algorithm 3), which is very similar to a sparse multiplication, but that can not be implemented using the PCLMULQDQ instruction. Moreover, our implementation relies only on instructions that are much more commonly available than a carry-less multiplier and that are also more likely to receive hardware improvements.

4   Results

First, in order to improve our analysis, we measured the use and cost of each of the arithmetic operations in QcBits [8]. Table 2 shows the results. We are using an updated version of QcBits [17] at the 128-bit quantum security level. The implementation is fully constant time and vectorized using AVX-512 instructions. Its constant-time decryption takes 11 iterations to achieve a decoding failure rate (DFR) smaller than 10^{-7} [17].

Table 2. Use and execution cost (cycles) of the arithmetic operations in QcBits at the 128-bit quantum security level.

Procedure	Addition			Multiplication			Inverse			Total
	#	Cost	%	#	Cost	%	#	Cost	%	
Key generation	0	0	0.00	2	260,046	0.65	1	39,762,389	98.75	40,265,904
Encryption	1	91	0.04	1	130,023	50.14	0	0	0.00	259,306
Decryption	44	4,004	0.04	45	5,851,035	59.68	0	0	0.00	9,803,835

As expected, the cost of the addition is negligible. The inversion takes 98.8% of the key generation and the multiplication takes 60% and 50% of the decryption and encryption, respectively. The remaining time in the key generation and encryption is taken by the random generation of the polynomials. In the decryption, the remaining time corresponds to the counting of satisfied parity-checks, represented by the function *IntegerPolynomialAddition* in Algorithm 3. To obtain a comparison basis for the operations, we executed two other inversion algorithms and one other multiplication algorithm in the same machine and for the same modulus polynomial. All of them use AVX-512 instructions and are highly optimized implementations. Table 3 presents the results.

For benchmarking, we have used version 11.3.0 of the NTL library [33] with support to the library gf2x for binary field arithmetic. Its inversion algorithm, used in BIKE, takes 12,088,846 cycles to invert, which is 1.7 times faster than our algorithm. Considering, however, that NTL's inversion is not constant-time, a 1.7 slowdown is still a good result for our algorithm, which is fully constant-time. The second evaluated algorithm for inversion was Itoh-Tsujii [20], one of the most used constant-time inversion algorithms. We implemented it using NTL's

Table 3. Comparison among implementations of multiplication and inversion. Bolded lines represent results from this paper.Comparison among implementations of multiplication and inversion. Bolded lines represent results from this paper.

Operation	Implementation	Failure rate	Constant time	Cost (cycles)
Inversion	**Wu *et al.* modified**	2^{-128}	Yes	20,773,925
	Wu *et al.* modified	10^{-8}	Yes	14,979,764
	Wu *et al.*	0	Yes	39,747,301
	NTL	0	No	12,088,846
	Itoh-Tsujii	0	Yes	243,226,595
Multiplication	**Sparse Mult.**	2^{-128}	Yes	79,843
	Sparse Mult.	0	Yes	130,023
	NTL	0	?	161,715

multiplications and squarings and one of the shortest addition chains [13]. It takes 243,226,595 cycles to invert a polynomial, which is 12 times slower than our implementation. For the multiplication, using NTL takes 161,715 cycles, which is more than 2 times slower than our multiplication algorithm, even considering the cost of sorting the polynomials (in the context of decryption).

In order to compute the impact of our algorithms in a real implementation, we introduced them in QcBits. Table 4 shows the execution time (cycles) for each of the procedures of the encryption scheme with and without the use of a negligible Failure Rate in the Arithmetic. We also present the execution time of BIKE (Variant 2) for comparison. We compiled the *Additional* version of BIKE (dated 05/23/2018) using NTL version 11.3.0, `gf2x` support and the following command: `make BIKE_VER=2 CONSTANT_TIME=1 RDTSC=1 AVX512=1 LEVEL=5`.

It is important to note that BIKE paper and this paper present different definitions for "constant-time implementations". In this paper, we define constant-time implementations as those in which the execution time does not depend on secret data. The authors of BIKE apparently define constant-time implementations as those which are not vulnerable to known timing side-channel attacks. Objectively, both definitions are sufficient to provide security against the currently known attacks, but they need to be differentiated to compare the performance results. For example, the "constant-time" decryption in BIKE executes a variable number of constant-time iterations. In this paper, we refer to it as a *Uniform Implementation*. The number of decoding iterations is dependent on the secret key and can be used to retrieve it [11]. However, all known attacks exploiting the number o decoding iterations rely on a large number of decoding attempts, whereas BIKE uses ephemeral keys [2]. Moreover, BIKE also makes use of masking and blinding techniques on non-constant time procedures, which we also do not consider as a constant-time implementation.

The speedups achieved in the Key Generation and in Decryption were the expected, considering the speedup of the operations and their use in the procedures. Encryption did not present any significant gains, despite having 50%

Table 4. Execution time (cycles) of QcBits, with (this paper) and without the Arithmetic Failure Rate; and of the *additional* implementation of BIKE-2, for comparison.

	128-bit QcBits			BIKE-2	
	This paper	Failure-less [17]	Speedup	Additional	Speedup
Key generation	21,332,058	40,265,904	1.89	12,944,920	0.61*
Encryption	256,655	259,306	1.01	348,227	1.36
Constant-time decrypt.	8,016,312	9,803,835	1.22	**	**
Uniform decryption	3,505,079	5,008,429	1.43	9,572,412	2.73

*BIKE's polynomial inversion is not constant-time.
**BIKE does not present constant-time decryption.
The results for BIKE are equivalent to the column "Constant time implementation" and line "AVX-512" from Table 19 in their paper [1].

of its execution time corresponding to a multiplication. This occurs due to the cost of sorting the monomials, as explained at the end of Sect. 3.2. The speedups over the official BIKE implementation are not a result solely of the techniques presented in this paper. As can be observed, the failure-less 128-bit version of QcBits [17] is already faster than BIKE. Nonetheless, the results help to support the advantages of our techniques, which can also be applied to BIKE. Our uniform decryption is 2.7 times faster than BIKE's and our key generation was 1.7 times slower. This slowdown is explained by the use of NTL's polynomial inversion, which is not constant-time.

5 Related Work

The use of imperfect algorithms as exchange for performance is vastly explored in many fields of computer science. One of the main examples is the field of Approximate Computing [7], which exploits imperfect algorithms to improve the performance of applications such as neural networks, image processing, encrypted data storage [21], among others. In the cryptography field, the main examples of imperfect algorithms targeting performance are probabilistic algorithms, such as the Miller-Rabin primality test [29], which is applied iteratively until the failure rate becomes negligible. Other failure rates present in the cryptography field are usually an undesired side effect from the explored underlying problem, which is the case of QC-MDPC code-based cryptography, for example. In its specific context, our approach differs from the previous QC-MDPC literature mainly for purposely introducing failures on arithmetic procedures that, otherwise, would return correct results.

Considering optimization techniques for QC-MDPC code-based cryptography, there are many works exploring techniques such as vectorization [23], bit-slicing [8], pre-computation [16], among others. Recently, Drucker and Gueron [10] presented several techniques for optimizing QC-MDPC implementations, which are currently used by the official BIKE implementation. Among them, a technique for optimizing the counting of unsatisfied parity-check bits in the bit flipping algorithm has a similar implementation to one of our algorithms. The authors inserted *"fake" bits* (monomials) in a polynomial before submitting it to a non-constant time procedure. After the procedure, they eliminate in constant time the results generated by the *"fake" bits*. This technique is very similar to the addition of an auxiliary polynomial in Algorithm 9 to help the multiplication and later discard the results generated by this polynomial. The differences reside on the objective and the generation of such auxiliary polynomial (or *"fake" bits*, in their nomenclature). Their bits were randomly generated and their objective was to implement a blinding technique for a non-constant time process. Meanwhile, our polynomial was systematically generated and our objective was to optimize a constant time process. Despite the similarities, we did not base our algorithm on their approach.

6   Conclusions

In this paper, we presented the concept of inserting a Failure Rate (FR) to the arithmetic of QC-MDPC code-based implementations to achieve significant performance gains. We provided algorithms for multiplication and inversion over binary polynomials that accept a controlled failure rate and defined methods to correlate the failure rate with the performance level. After introducing a negligible FR in these algorithms, we achieved a 1.9-factor performance speedup in the inversion and a 1.63-factor performance speedup in the multiplication. We also showed that our multiplication is 2 times faster than NTL's and that our inversion is 12 times faster than Itoh-Tsujii. Finally, we used our algorithms in the QcBits implementation, where we achieved speedups of 1.89 and up to 1.43 in the key generation and decryption, respectively.

Our experimental results show the performance impact of our approach, while the negligible failure rate has basically no downsides to the cryptosystem. The correlation between failure rate and performance improvement was also shown to be very promising, once it is possible to achieve much lower failure rates with little performance penalties. The algorithms we presented have certain advantages when used in QC-MDPC cryptosystems, but, ultimately, they are generic algorithms for arithmetic in $GF(2^n)$ and, thus, could be used in other contexts.

As future work, we intend to insert and benchmark these algorithms in BIKE and to pursue further performance improvements. We also intend to replicate the approach to other expensive algorithms used in QC-MDPC cryptosystems or other arithmetic algorithms of cryptographic interest. Examples of promising sub-procedures in QC-MDPC cryptosystems for the introduction of failures are the Unsatisfied Parity-Check Counting (line 9 of Algorithm 3) and the syndrome update, which can be used to replace the syndrome recalculation (line 13 of Algorithm 3).

Acknowledgements. We would like to thank CNPq, Intel, and the São Paulo Research Foundation (FAPESP) for supporting this research under grants 313012/2017-2, 2014/50704-7 and 2013/08293-7. We also would like to thank Microsoft for providing the cloud infrastructure for our experiments with state-of-the-art microprocessors. And, finally, we would like to thank Professor Julio López for the useful comments and ideas.

A Implementing Conditional Statements (Ifs) in Constant-Time

Listing 1 shows an example of a non-constant-time conditional operation. Assuming that A and B are secret data, this implementation is vulnerable to timing side-channel attacks. Listing 2 shows the equivalent constant-time implementation (considering that Function1 and Function2 do not have side effects). The operations using 64-bit integers (uint64_t) had their results conditionally selected through a bit-wise AND with the mask cond. When using AVX-512 registers, the implementation of conditional operations is significantly simplified. The AVX-512 instruction set extension already provides masked versions for most of its instructions. In this way, we simply use the mask cond in the mask field of the intrinsics of these instructions.

```
uint64_t A, B, C, D, cond;
__m512i V1, V2, V3;

[...]

if(A < B){
    C = Function1();
    D += 5;
    V1 -= V2;
}else{
    C = Function2();
    D ^= 0xf;
    V1 &= V3;
}
```

Listing 1: Non-constant-time conditional operations

```
cond = ((int64_t) (A - B)) >> (63);

C = cond & Function1() | ~cond &  Function2();
D = cond & (D + 5) | ~cond & (D ^ 0xf);

V1 = _mm512_mask_sub_epi64(V1, cond, V1, V2);
V1 = _mm512_mask_and_epi64(V1, ~cond, V1, V3);
```

Listing 2: Constant-time conditional operations

B   Implementing the Degree Verification Efficiently

Our modification of Wu *et al.* algorithm introduced two main drawbacks in the performance of the algorithm. The first is the constant-time implementation of the function *Smallest_Monomial_Degree*. For large polynomials (such as the ones used in code-based cryptography), it would be very expensive to search the smallest monomial on the entire polynomial. Therefore, we search only the first E bits of the polynomial, change the If condition to test if the result is different of E and adjust the number of iterations to compensate for this limitation. Algorithm 11 shows this modification. Using the inverted binary representation of the polynomial (shown in Fig. 5), we can obtain the degree of the smallest monomial by calculating the number of leading zeros of the representation. Most of the modern architectures enable this calculation with a single instruction. Intel, for instance, provides instructions for counting the leading zeros running in constant-time for 32-bit words (since i386), 64-bit words (since Haswell), and 64-bit lanes on SIMD registers (AVX-512). Other architectures enable equivalent or complementary operations, such as rounded binary logarithm or trailing zeros, which may require modifications in the polynomial representation, but, ultimately, would not impact performance.

$$x^7 + x^3 + x^2 + 1 \leftrightarrow [1, 0, 1, 1, 0, 0, 0, 1] \leftrightarrow [0, 2, 3, 7]$$

Fig. 5. Example polynomial, its inverted binary representation and its sparse representation, respectively.

Algorithm 11: Modified version of Wu *et al.* Inversion Algorithm [37] considering the parameter E.

Input : r, s, u, N, F and E
Output: $v = \frac{u}{r} \pmod{s}$
1 $v \leftarrow 0$, $\delta \leftarrow -1$, $g \leftarrow s$
2 **for** $i = 0$ **to** $F \times (2 \times N + \frac{2 \times N}{E})$ **do**
3 $\quad$ $b \leftarrow Smallest\_Monomial\_Degree(r, E)$
4 $\quad$ $(r, u) \leftarrow (r/x^b, (u/x^b)_g)$
5 $\quad$ $\delta \leftarrow \delta - b$
6 $\quad$ **if** $b \neq E$ **then**
7 $\quad\quad$ | *If's content, unchanged*

The second drawback in our version are the divisions. In the original algorithm, the divisor was always x. We modified it to x^b, where $0 < b \leq E$. Constant-time divisions usually have its execution time defined by the upper bound of the divisor and, thus, the parameter E also appears as a trade-off between the number of iterations and the performance of each iteration. Fortunately, it is easy to optimize its value in our case. Using SIMD registers in the Intel architecture, the execution time of dividing by x or x^{64} is the same, while greater exponents require much more expensive instructions to move bits across the SIMD lanes. In this way, we choose $E = 64$, which also helps the implementation of the function $Smallest\_Monomial\_Degree$.

C   Proof of Eq. 2

Proposition 1. *In Eq. 2, if $P(r_{j,0} = 1) \leq 0.5$ then $P(r_{j,i} = 1) \leq 0.5$ for all $i \geq 0$.*

Proof. We prove it using induction on i.
Base case: If $i = 0$, then $P(r_{j,i} = 1) = P(r_{j,0} = 1) \leq 0.5$.
Inductive Hypothesis: if $P(r_{j,0} = 1) \leq 0.5$ then $P(r_{j,i} = 1) \leq 0.5$ for all $1 \leq i < n$
Inductive Step: Let $i = n$.

$$P(r_{j,n} = 1) = P(r_{0,n-1} = 1) \times P(s_{j+1,n-1} \oplus r_{j+1,n-1} = 1) + P(r_{0,n-1} = 0)$$
$$\times P(r_{j+1,n-1} = 1)$$

Knowing that

$$P((X \oplus Y) = 1) = P(X = 1) \times P(Y = 0) + P(X = 0) \times P(Y = 1)$$
$$= P(X = 1) \times (1 - P(Y = 1)) + (1 - P(X = 1)) \times P(Y = 1)$$
$$= P(X = 1) + P(Y = 1) - 2 \times P(Y = 1) \times P(X = 1)$$

We have

$$P(r_{j,n} = 1) = P(r_{0,n-1} = 1) \times [P(r_{j+1,n-1} = 1) + P(s_{j+1,n-1} = 1)$$
$$-2 \times P(s_{j+1,n-1} = 1) \times P(r_{j+1,n-1} = 1)] + P(r_{0,n-1} = 0) \times P(r_{j+1,n-1} = 1)$$
$$= P(r_{j+1,n-1} = 1) + P(r_{0,n-1} = 1) \times [P(s_{j+1,n-1} = 1)$$
$$-2 \times P(s_{j+1,n-1} = 1) \times P(r_{j+1,n-1} = 1)]$$

Let $f(X, Y, Z)$ be the value of $P(r_{j,n} = 1)$ for $X = P(r_{0,n-1} = 1)$, $Y = P(r_{j+1,n-1} = 1)$ and $Z = P(s_{j+1,n-1} = 1)$. By our inductive hypothesis, $0 \leq P(r_{0,n-1} = 1) \leq 0.5$ and $0 \leq P(r_{j+1,n-1} = 1) \leq 0.5$. We could obtain a tighter interval for $P(s_{j+1,n-1} = 1)$ addressing its own recurrence relation, but that is not necessary. Thus, we consider $0 \leq P(s_{j+1,n-1} = 1) \leq 1$. To maximize the value of f in these intervals, we first check the boundaries:

$$f(0,0,0) = f(0,0,1) = f(0.5,0,0) = 0$$

$$f(0,0.5,0) = f(0,0.5,1) = f(0.5,0,1) = f(0.5,0.5,0) = f(0.5,0.5,1) = 0.5$$

The next step would be a search for a local maximum, which clearly does not exist since f is linear in all variables.

References

1. Aragon, N., et al.: BIKE: bit flipping key encapsulation, December 2017. https://hal.archives-ouvertes.fr/hal-01671903. Submission to the NIST post quantum standardization process. Website: http://bikesuite.org/
2. Barreto, P.S.L.M., et al.: Cake: code-based algorithm for key encapsulation. Cryptology ePrint Archive, Report 2017/757 (2017). http://eprint.iacr.org/2017/757
3. Berlekamp, E., McEliece, R., van Tilborg, H.: On the inherent intractability of certain coding problems (corresp.). IEEE Trans. Inf. Theory **24**(3), 384–386 (1978). https://doi.org/10.1109/TIT.1978.1055873
4. Bernstein, D.J.: SUPERCOP: system for unified performance evaluation related to cryptographic operations and primitives (2009)
5. Bernstein, D.J., Chuengsatiansup, C., Lange, T., van Vredendaal, C.: NTRU prime: reducing attack surface at low cost. In: Adams, C., Camenisch, J. (eds.) SAC 2017. LNCS, vol. 10719, pp. 235–260. Springer, Cham (2018). https://doi.org/10.1007/978-3-319-72565-9_12
6. Brunner, H., Curiger, A., Hofstetter, M.: On computing multiplicative inverses in $GF(2^m)$. IEEE Trans. Comput. **42**(8), 1010–1015 (1993). https://doi.org/10.1109/12.238496
7. Ceze, L., et al.: Disciplined approximate computing: from language to hardware, and beyond. Personal Web-page, https://homes.cs.washington.edu/~luisceze/ceze-approx-overview.pdf
8. Chou, T.: QcBits: constant-time small-key code-based cryptography. In: Gierlichs, B., Poschmann, A.Y. (eds.) CHES 2016. LNCS, vol. 9813, pp. 280–300. Springer, Heidelberg (2016). https://doi.org/10.1007/978-3-662-53140-2_14
9. Drucker, N., Gueron, S., Krasnov, V.: Fast multiplication of binary polynomials with the forthcoming vectorized VPCLMULQDQ instruction. In: 2018 IEEE 25th Symposium on Computer Arithmetic (ARITH), pp. 115–119, June 2018. https://doi.org/10.1109/ARITH.2018.8464777
10. Drucker, N., Gueron, S.: A toolbox for software optimization of QC-MDPC code-based cryptosystems. J. Cryptogr. Eng. (2019). https://doi.org/10.1007/s13389-018-00200-4
11. Eaton, E., Lequesne, M., Parent, A., Sendrier, N.: QC-MDPC: a timing attack and a CCA2 KEM. In: Lange, T., Steinwandt, R. (eds.) PQCrypto 2018. LNCS, vol. 10786, pp. 47–76. Springer, Cham (2018). https://doi.org/10.1007/978-3-319-79063-3_3
12. Faugère, J.-C., Otmani, A., Perret, L., Tillich, J.-P.: Algebraic cryptanalysis of McEliece variants with compact keys. In: Gilbert, H. (ed.) EUROCRYPT 2010. LNCS, vol. 6110, pp. 279–298. Springer, Heidelberg (2010). https://doi.org/10.1007/978-3-642-13190-5_14
13. Flammenkamp, A.: Shortest addition chains. Achim's WWW Domain (2018). http://wwwhomes.uni-bielefeld.de/achim/addition_chain.html
14. Gallager, R.: Low-Density Parity-Check Codes. MIT press, Cambridge (1963)
15. Goppa, V.D.: A new class of linear correcting codes. Problemy Peredachi Informatsii **6**(3), 24–30 (1970)
16. Guimarães, A., Aranha, D.F., Borin, E.: Optimizing the decoding process of a post-quantum cryptographic algorithm. In: XVIII Simpósio em Sistemas Computacionais de Alto Desempenho-WSCAD, vol. 18, no. 1/2017, pp. 160–171 (2017)
17. Guimarães, A., Aranha, D.F., Borin, E.: Optimized implementation of QC-MDPC code-based cryptography. Concurr. Comput. Pract. Exp. (2018). https://doi.org/10.1002/cpe.5089

18. Hamming, R.W.: Coding and Information Theory, 2nd edn. Prentice-Hall Inc., Upper Saddle River (1986)
19. Hankerson, D., Menezes, A.J., Vanstone, S.: Guide to Elliptic Curve Cryptography. SPC. Springer, New York (2004). https://doi.org/10.1007/B97644
20. Itoh, T., Tsujii, S.: A fast algorithm for computing multiplicative inverses in $GF(2^m)$ using normal bases. Inf. Comput. **78**(3), 171–177 (1988). https://doi.org/10.1016/0890-5401(88)90024-7
21. Jevdjic, D., Strauss, K., Ceze, L., Malvar, H.S.: Approximate storage of compressed and encrypted videos, vol. 45, pp. 361–373. ACM, New York, April 2017. https://doi.org/10.1145/3093337.3037718
22. Kou, Y., Xu, J., Tang, H., Lin, S., Abdel-Ghaffar, K.: On circulant low density parity check codes. In: Proceedings IEEE International Symposium on Information Theory, p. 200 (2002). https://doi.org/10.1109/ISIT.2002.1023472
23. Maurich, I.V., Oder, T., Güneysu, T.: Implementing QC-MDPC McEliece encryption. ACM Trans. Embed. Comput. Syst. **14**(3), 44:1–44:27 (2015). https://doi.org/10.1145/2700102
24. McEliece, R.J.: A public-key cryptosystem based on algebraic coding theory. Deep Space Network Progress Report 44, pp. 114–116 (1978)
25. Misoczki, R., Tillich, J.P., Sendrier, N., Barreto, P.S.L.M.: MDPC-McEliece: new McEliece variants from moderate density parity-check codes. In: 2013 IEEE International Symposium on Information Theory, pp. 2069–2073, July 2013. https://doi.org/10.1109/ISIT.2013.6620590
26. Monico, C., Rosenthal, J., Shokrollahi, A.: Using low density parity check codes in the McEliece cryptosystem. In: 2000 IEEE International Symposium on Information Theory, p. 215 (2000). https://doi.org/10.1109/ISIT.2000.866513
27. Niederreiter, H.: Knapsack-type cryptosystems and algebraic coding theory. Prob. Control Inf. Theory **15**(2), 159–166 (1986)
28. NIST: Submission requirements and evaluation criteria for the post-quantum cryptography standardization process. NIST web page (2016). http://csrc.nist.gov/groups/ST/post-quantum-crypto/documents/call-for-proposals-final-dec-2016.pdf
29. Rabin, M.O.: Probabilistic algorithm for testing primality. J. Number Theory **12**(1), 128–138 (1980). https://doi.org/10.1016/0022-314X(80)90084-0
30. Rivest, R.L., Shamir, A., Adleman, L.: A method for obtaining digital signatures and public-key cryptosystems. Commun. ACM **21**(2), 120–126 (1978). https://doi.org/10.1145/359340.359342
31. Rossi, M., Hamburg, M., Hutter, M., Marson, M.E.: A side-channel assisted cryptanalytic attack against QcBits. In: Fischer, W., Homma, N. (eds.) CHES 2017. LNCS, vol. 10529, pp. 3–23. Springer, Cham (2017). https://doi.org/10.1007/978-3-319-66787-4_1
32. Shor, P.W.: Polynomial-time algorithms for prime factorization and discrete logarithms on a quantum computer. SIAM J. Comput. **26**(5), 1484–1509 (1997). https://doi.org/10.1137/S0097539795293172
33. Shoup, V.: Number Theory C++ Library (NTL) (2003)
34. Stanley, R.P.: What Is Enumerative Combinatorics?. The Wadsworth & Brooks/Cole Mathematics Series, vol. 1, pp. 1–63. Springer, Boston (1986). https://doi.org/10.1007/978-1-4615-9763-6_1
35. Stein, J.: Computational problems associated with Racah algebra. J. Comput. Phys. **1**(3), 397–405 (1967). https://doi.org/10.1016/0021-9991(67)90047-2

36. Strenzke, F., Tews, E., Molter, H.G., Overbeck, R., Shoufan, A.: Side channels in the McEliece PKC. In: Buchmann, J., Ding, J. (eds.) PQCrypto 2008. LNCS, vol. 5299, pp. 216–229. Springer, Heidelberg (2008). https://doi.org/10.1007/978-3-540-88403-3_15
37. Wu, C.H., Wu, C.M., Shieh, M.D., Hwang, Y.T.: High-speed, low-complexity systolic designs of novel iterative division algorithms in $GF(2^m)$. IEEE Trans. Comput. **53**(3), 375–380 (2004). https://doi.org/10.1109/TC.2004.1261843

DAGS: Reloaded Revisiting Dyadic Key Encapsulation

Gustavo Banegas[1], Paulo S. L. M. Barreto[2], Brice Odilon Boidje[3],
Pierre-Louis Cayrel[4], Gilbert Ndollane Dione[3], Kris Gaj[5],
Cheikh Thiécoumba Gueye[3], Richard Haeussler[5], Jean Belo Klamti[3],
Ousmane N'diaye[3], Duc Tri Nguyen[5], Edoardo Persichetti[6(✉)],
and Jefferson E. Ricardini[7]

[1] Tecnische Universiteit Eindhoven, Eindhoven, The Netherlands
[2] University of Washington Tacoma, Tacoma, USA
[3] Laboratoire d'Algebre, de Cryptographie, de Géométrie Algébrique et Applications,
Université Cheikh Anta Diop, Dakar, Senegal
[4] Laboratoire Hubert Curien, Université Jean Monnet, Saint-Etienne, France
[5] George Mason University, Fairfax, USA
[6] Department of Mathematical Sciences,
Florida Atlantic University, Boca Raton, USA
epersichetti@fau.edu
[7] Universidade de São Paulo, São Paulo, Brazil

Abstract. In this paper we revisit some of the main aspects of the DAGS Key Encapsulation Mechanism, one of the code-based candidates to NIST's standardization call for the key exchange/encryption functionalities. In particular, we modify the algorithms for key generation, encapsulation and decapsulation to fit an alternative KEM framework, and we present a new set of parameters that use binary codes. We discuss advantages and disadvantages for each of the variants proposed.

Keywords: Post-quantum cryptography · Code-based cryptography · Key exchange

1 Introduction

The majority of cryptographic protocols in use in the present day are based on traditional problems from number theory such as factoring or computing discrete logarithms in some group; this is the case for schemes such as RSA, DSA, ECDSA etc. This is undoubtedly about to change due to the looming threat of quantum computers. In fact, due to Shor's seminal work [21], such problems will be vulnerable to polynomial-time attacks once quantum computers with enough computational power are available, which will make current

© Springer Nature Switzerland AG 2019
M. Baldi et al. (Eds.): CBC 2019, LNCS 11666, pp. 69–85, 2019.
https://doi.org/10.1007/978-3-030-25922-8_4

cryptographic solutions obsolete. While the resources necessary to effectively run Shor's algorithm on actual cryptographic parameters (or other cryptographically relevant quantum algorithms such as or Grover's [12]) might be at least a decade away, post-quantum cryptography cannot wait for this to happen. In fact, today's encrypted communication could be easily stored by attackers and decrypted later with a quantum computer, compromising secrets that aim for long-term security. Therefore, it is vital that the time required to develop such resources (t_{Dev}) is not inferior to the sum of the time required to develop and deploy new cryptographic standards (t_{Dep}), and the desired lifetime of a secret (t_{Sec}), i.e. we need to ensure $t_{Dev} \geq t_{Dep} + t_{Sec}$.

With this in mind, the National Institute of Standards and Technology (NIST) has launched a Call for Proposals for Post Quantum Cryptographic Schemes [17], to select a range of post-quantum primitives to become the new public-key cryptography standard. The NIST Call is soliciting proposals in encryption, key exchange, and digital signature schemes. It is expected that the effort will require approximately 5 years, and another 5 years will likely follow for the deployment phase, which includes developing efficient implementations and updating the major cryptographic products to the new standard. Currently, there are five major families of post-quantum cryptosystems: lattice-based, code-based, hash-based, isogeny-based, and multivariate polynomial-based systems. The first two are the most investigated, comprising nearly three quarters of the total amount of submissions to the NIST competition (45 out of 69).

Our Contribution. DAGS [1] was one of the candidates to NIST's Post-Quantum Standardization call [17]. The submission presents a code-based Key Encapsulation Mechanism (KEM) that uses Quasi-Dyadic (QD) Generalized Srivastava (GS) codes to achieve very small sizes for all the data (public and private key, and ciphertext); as a result, DAGS features one of the smallest data size among all the code-based submissions. Unfortunately, due to some security concerns, DAGS was not selected for the second round of the standardization call. In this paper, we present the results of our investigation of the DAGS scheme, aimed at tweaking and improving several aspects of the scheme. First, we describe a new approach to the protocol design (Sect. 3), which offers an important alternative and a tradeoff between security and performance. Then, in Sect. 4, we propose and discuss new parameters, including an all-new set based on binary codes, to protect against both known and new attacks. Finally, in Sect. 5, we report the numbers obtained in a new, improved implementation which uses dedicated techniques and tricks to achieve a considerable speed up. We hope this version of DAGS can provide some stability and reassurance about the security of the system.

2   Preliminaries

2.1   Notation

We will use the following conventions throughout the rest of the paper (Table 1):

Table 1. Notation used in this document.

a	A constant
$\boldsymbol{a}$	A vector
A	A matrix
$\mathcal{A}$	An algorithm or (hash) function
A	A set
$(\boldsymbol{a} \parallel \boldsymbol{b})$	The concatenation of vectors $\boldsymbol{a}$ and $\boldsymbol{b}$
$\mathrm{Diag}(\boldsymbol{a})$	The diagonal matrix formed by the vector $\boldsymbol{a}$
I_n	The $n \times n$ identity matrix
$\xleftarrow{\$}$	Choosing a random element from a set or distribution
ℓ	The length of a shared symmetric key

2.2 Linear Codes

We briefly recall some fundamental notions from Coding Theory. The *Hamming weight* of a vector $\boldsymbol{x} \in \mathbb{F}_q^n$ is given by the number $\mathsf{wt}(\boldsymbol{x})$ of its nonzero components. We define a linear code using the metric induced by the Hamming weight.

Definition 1. *An $[n, k]$-linear code C of length n and dimension k over $\mathbb{F}_q$ is a k-dimensional vector subspace of $\mathbb{F}_q^n$.*

A linear code can be represented by a matrix $G \in \mathbb{F}_q^{k \times n}$, called *generator matrix*, whose rows form a basis for the vector space defining the code. Alternatively, a linear code can also be represented as kernel of a matrix $H \in \mathbb{F}_q^{(n-k) \times n}$, known as *parity-check matrix*, i.e. $\mathsf{C} = \{\boldsymbol{c} : H\boldsymbol{c}^T = 0\}$. Thanks to the generator matrix, we can easily define the codeword corresponding to a vector $\boldsymbol{\mu} \in \mathbb{F}_q^k$ as $\boldsymbol{\mu}G$. Finally, we call *syndrome* of a vector $\boldsymbol{c} \in \mathbb{F}_q^n$ the vector $H\boldsymbol{c}^T$.

2.3 Structured Matrices and GS Codes

Definition 2. *Given a ring R (in our case the finite field $\mathbb{F}_{q^m}$) and a vector $\boldsymbol{h} = (h_0, \ldots, h_{n-1}) \in \mathsf{R}^n$, the dyadic matrix $\Delta(\boldsymbol{h}) \in \mathsf{R}^{n \times n}$ is the symmetric matrix with components $\Delta_{ij} = h_{i \oplus j}$, where $\oplus$ stands for bitwise exclusive-or on the binary representations of the indices. The sequence $\boldsymbol{h}$ is called its signature. Moreover, $\Delta(r, \boldsymbol{h})$ denotes the matrix $\Delta(\boldsymbol{h})$ truncated to its first r rows. Finally, we call a matrix quasi-dyadic if it is a block matrix whose component blocks are $r \times r$ dyadic submatrices.*

If n is a power of 2, then every $2^l \times 2^l$ dyadic matrix can be described recursively as

$$M = \begin{pmatrix} A & B \\ B & A \end{pmatrix}$$

where each block is a $2^{l-1} \times 2^{l-1}$ dyadic matrix. Note that by definition any 1 × 1 matrix is trivially dyadic.

Definition 3. *For $m, n, s, t \in \mathbb{N}$ and a prime power q, let $\alpha_1, \ldots, \alpha_n$ and $w_1, \ldots, w_s$ be $n + s$ distinct elements of $\mathbb{F}_{q^m}$, and $z_1, \ldots, z_n$ be nonzero elements of $\mathbb{F}_{q^m}$. The Generalized Srivastava (GS) code of order st and length n is defined by a parity-check matrix of the form:*

$$H = \begin{pmatrix} H_1 \\ H_2 \\ \vdots \\ H_s \end{pmatrix}$$

where each block is given by

$$H_i = \begin{pmatrix} \dfrac{z_1}{\alpha_1 - w_i} & \cdots & \dfrac{z_n}{\alpha_n - w_i} \\[2mm] \dfrac{z_1}{(\alpha_1 - w_i)^2} & \cdots & \dfrac{z_n}{(\alpha_n - w_i)^2} \\[1mm] \vdots & \vdots & \vdots \\[1mm] \dfrac{z_1}{(\alpha_1 - w_i)^t} & \cdots & \dfrac{z_n}{(\alpha_n - w_i)^t} \end{pmatrix}.$$

The parameters for such a code are the length $n \leq q^m - s$, dimension $k \geq n - mst$ and minimum distance $d \geq st + 1$. GS codes are part of the family of alternant codes, and therefore benefit of an efficient decoding algorithm; according to Sarwate [20, Cor. 2] the complexity of decoding is $\mathcal{O}(n \log^2 n)$, which is the same as for Goppa codes. Moreover, it can be easily proved that every GS code with $t = 1$ is a Goppa code. More information about this class of codes can be found in [15, Ch. 12, §6].

3   SimpleDAGS

3.1   Construction

The core idea of DAGS is to use GS codes which are defined by matrices in quasi-dyadic form. It can be easily proved that every GS code with $t = 1$ is a Goppa code, and we know [15, Ch. 12, Pr. 5] that Goppa codes admit a parity-check matrix in Cauchy form under certain conditions (the generator polynomial has to be monic and without multiple zeros). By Cauchy we mean a matrix $C(\boldsymbol{u}, \boldsymbol{v})$ with components $C_{ij} = \frac{1}{u_i - v_j}$. Misoczki and Barreto showed in [16, Th. 2] that the intersection of the set of Cauchy matrices with the set of dyadic matrices is not empty if the code is defined over a field of characteristic 2, and the dyadic signature $\boldsymbol{h} = (h_0, \ldots, h_{n-1})$ satisfies the following "fundamental" equation

$$\frac{1}{h_{i \oplus j}} = \frac{1}{h_i} + \frac{1}{h_j} + \frac{1}{h_0}. \tag{1}$$

On the other hand, it is evident from Definition 3 that if we permute the rows of H to constitute $s \times n$ blocks of the form

$$\hat{H}_i = \begin{pmatrix} \dfrac{z_1}{(\alpha_1 - w_1)^i} & \cdots & \dfrac{z_n}{(\alpha_n - w_1)^i} \\ \dfrac{z_1}{(\alpha_1 - w_2)^i} & \cdots & \dfrac{z_n}{(\alpha_n - w_2)^i} \\ \vdots & \vdots & \vdots \\ \dfrac{z_1}{(\alpha_1 - w_s)^i} & \cdots & \dfrac{z_n}{(\alpha_n - w_s)^i} \end{pmatrix}$$

we obtain an equivalent parity-check matrix for a GS code, given by

$$\hat{H} = \begin{pmatrix} \hat{H}_1 \\ \hat{H}_2 \\ \vdots \\ \hat{H}_t \end{pmatrix}.$$

The key generation process exploits first of all the fundamental equation to build a Cauchy matrix. The matrix is then successively powered (element by element) forming several blocks which are superimposed and then multiplied by a random diagonal matrix. Thanks to the observation above, we have now formed the matrix $\hat{H}$, where for ease of notation we use $\boldsymbol{u}$ and $\boldsymbol{v}$ to denote the vectors of elements $w_1, \ldots, w_s$ and $\alpha_1, \ldots, \alpha_n$, respectively. Finally, the resulting matrix is projected onto the base field (as usual for alternant codes) and row-reduced to systematic form to form the public key. The process is essentially the same as in [19], to which we refer the readers looking for additional details about dyadic GS codes and key generation.

3.2 The New Algorithms

The DAGS algorithms, as detailed in the original proposal submitted to the first Round [8], follow the "Randomized McEliece" paradigm of Nojima et al. [18], which is built upon the McEliece cryptosystem. The fact that this version of McEliece is proved to be IND-CPA secure makes it so that the resulting KEM conversion achieves IND-CCA security tightly, as detailed in [13]. However, to apply the conversion correctly, it is necessary to use multiple random oracles. These are needed to produce the additional randomness required by the paradigm, as well as to convert McEliece into a deterministic scheme (by generating a low-weight error vector from a random seed) and to obtain an additional hash output for the purpose of plaintext confirmation. Even though, in practice, such random oracles are realized using the same hash function (the *KangarooTwelve* function [14] from the Keccak family), the protocol's description ends up being quite cumbersome and hard to follow.

A simpler protocol can be obtained, although, as we will see, not without consequences, using the Niederreiter cryptosystem. We report the new description below, following the same conventions used in the original DAGS specification [1]. Therefore, system parameters are the code length n, dimension k and co-dimension r, the values s and t which define a GS code, the cardinality of the base field q and the degree of the field extension m. Note that, unlike the original DAGS, we do not need the sub-parameters k' and k''.

Algorithm 1. Key Generation

Key generation follows closely the process described in the original DAGS Key Generation. We present here a compact version, and we refer the reader to the description in Sect. 3.1.1 of [8] for further details.

1. Generate dyadic signature h.
2. Build the Cauchy support (u, v).
3. Form Cauchy matrix $\hat{H}_1 = C(u, v)$.
4. Build $\hat{H}_i$, $i = 2, \ldots t$, by raising each element of $\hat{H}_1$ to the power of i.
5. Superimpose blocks $\hat{H}_i$ in ascending order to form matrix $\hat{H}$.
6. Generate vector z by sampling uniformly at random elements in $\mathbb{F}_{q^m}$ with the restriction $z_{is+j} = z_{is}$ for $i = 0, \ldots, n_0 - 1$, $j = 0, \ldots, s - 1$.
7. Form $H = \hat{H} \cdot \mathrm{Diag}(z)$.
8. Project H onto $\mathbb{F}_q$ using the co-trace function: call this H_{base}.
9. Write H_{base} as $(B \mid A)$, where A is $r \times r$.
10. Get systematic form[1] $(M \mid I_r) = A^{-1} H_{base}$: call this $\tilde{H}$.
11. Sample a uniform random string $r \in \mathbb{F}_q^n$.
12. The public key is the matrix $\tilde{H}$.
13. The private key consists of (u, A, r) and $\tilde{H}$.

The main differences are as follows. First of all, the public key consists of the systematic parity-check matrix $\tilde{H} = (M \mid I_r)$, rather than the generator matrix $G = (I_k \mid M^T)$. Also, the private key only stores u instead of v and y, but it includes additional elements, namely the random string r, the submatrix A and $\tilde{H}$ itself[2].

---

[1] If A is not invertible, abort and go back to 1.

[2] This is mostly a formal difference, since $\tilde{H}$ is in fact the public key.

Algorithm 2. Encapsulation

Encapsulation uses a hash function $\mathcal{H} : \{0,1\}^* \rightarrow \{0,1\}^\ell$ to extract the desired symmetric key, ℓ being the desired bit length (commonly 256). The function is also used to provide plaintext confirmation by appending an additional hash value, as detailed below.

1. Sample $e \xleftarrow{\$} \mathbb{F}_q^n$ of weight w.
2. Set $c = (c_0, c_1)$ where $c_0 = \tilde{H}e$ and $c_1 = \mathcal{H}(2, e)$.
3. Compute $k = \mathcal{H}(1, e, c)$.
4. Output ciphertext c; the encapsulated key is k.

Algorithm 3. Decapsulation

As in every code-based scheme, the decapsulation algorithm consists mainly of decoding; in this case, like in the original DAGS version, we call upon the alternant decoding algorithm (see for example [15]).

1. Get syndrome c_0' corresponding to matrix[3] H' from private key[4].

2. Decode c_0 and obtain e'.

3. If decoding fails or $\mathsf{wt}(e') \neq w$, set $b = 0$ and $\eta = r$.

4. Check that $\tilde{H}e' = c_0$ and $\mathcal{H}(2, e') = c_1$. If so, set $b = 1$ and $\eta = e'$.

5. Otherwise, set $b = 0$ and $\eta = r$.

6. The decapsulated key is $k = \mathcal{H}(b, \eta, c)$.

The description we just presented is conform to the guidelines detailed by the "SimpleKEM" construction of [5], hence our choice to call this new version "SimpleDAGS". This is one of two aspects in which this variant diverges substantially from the original; we will discuss advantages (and disadvantages) of this new paradigm in the next section. The other different aspect is that using Niederreiter requires a different strategy for decoding, which we describe below.

3.3 Decoding from a Syndrome

In the original version of DAGS, the input to the decoding algorithm is, as is commonly the case is coding theory, a noisy codeword. The alternant decoding algorithm consists of three distinct steps. First, it is necessary to compute the syndrome of the received word, with respect to the alternant parity-check matrix; this is represented as a polynomial $S(z)$. Then, the algorithm uses the syndrome to compute the *error locator polynomial* $\sigma(z)$ and the *error evaluator polynomial* $\omega(z)$, by solving the *key equation* $\omega(z)/\sigma(z) = S(z) \mod z^r$. Finally, finding the roots of the two polynomials reveals, respectivelly, the locations and values (if the code is not binary) of the errors. Actually, as shown in Sect. 6.3 of [1], it is

---

[3] In alternant form.
[4] See next section for details.

possible to speed up decapsulation by incorporating the first step of the decoding algorithm in the reconstruction of the alternant matrix, i.e. the syndrome is computed *on the fly*, while the alternant matrix is built.

We are now ready to explain how to perform alternant decoding when the input is a syndrome, rather than a noisy codeword, as is the case in Algorithm 3 above. In this case, we do not need to reconstruct the alternant matrix itself, but rather to transform the received syndrome to the syndrome corresponding to the alternant matrix. This consists of two steps. First, remember that the public key $\tilde{H}$ is the systematic form of the matrix H_{base}. This is obtained from the quasi-dyadic parity-check matrix H, whose entries are in $\mathbb{F}_{q^m}$, by projecting it onto the base field $\mathbb{F}_q$. The projection is performed using the co-trace function and a basis for the extension field, say $\{\beta_1, \ldots, \beta_m\}$. Recall that the co-trace function works similarly to the trace function, by writing each element of $\mathbb{F}_{q^m}$ as a vector whose components are the coefficients with respect to the basis $\{\beta_1, \ldots, \beta_m\}$. However, instead of writing the components on m successive rows, the co-trace function distributes them over the rows at regular intervals, r at a time. More precisely, if $a = (a_1, a_2, \ldots, a_r)^T$ is a column of H, the corresponding column $a' = (a'_1, a'_2, \ldots, a'_{rm})^T$ of H_{base} will be formed by writing the components of each a_i in positions $a'_i, a'_{r+i}, \ldots, a'_{r(m-1)+i}$, for all $i = 1, \ldots, m$.

The first step consists of transforming the received syndrome $c_0 = \tilde{H}e$ into He. For this, we need to multiply the syndrome by A to obtain $A\tilde{H}e = AA^{-1}H_{base}e = H_{base}e$. Then we reverse the projection process and "bring back" the syndrome on the extension field. This is immediate when operating directly on the matrices, but a little less intuitive when starting from a syndrome. It turns out that it is still possible to do that, by using again the basis $\{\beta_1, \ldots, \beta_m\}$. Namely, it is enough to collect all the components $s_i, s_{r+i}, \ldots, s_{r(m-1)+i}$ of the syndrome $s = H_{base}e$ and multiply the resulting vector with the vector $(\beta_1, \ldots, \beta_m)$. This maps the vector of components back to its corresponding element in $\mathbb{F}_{q^m}$ and it is immediate to check that this process yields He.

The second step consists of relating the newly-obtained syndrome to the alternant parity-check matrix H'. Since this is just another parity-check for the same code, it is possible to obtain one from the other via an invertible matrix. In particular, for GS codes we have $H = CH'$, where the $r \times r$ matrix C can be obtained using u. Namely, the r rows of C corresponds to the coefficients of the polynomials $g_1(x), \ldots, g_r(x)$, where we have

$$g_{(l-1)t+i} = \frac{\prod_{j=1}^{s}(x - u_j)^t}{(x - u_l)^i}$$

for $l = 1, \ldots, s$ and $i = 1, \ldots, t$. To complete the second step, then, it is enough to compute C and then $C^{-1}He$. The resulting syndrome is ready to be decoded.

3.4   Consequences

There are some notable consequence to keep in mind when switching to the SimpleDAGS variant. First of all, the change in the KEM conversion not only makes the protocol simpler, but has additional advantages. The reduction is tight in the ROM, and the introduction of the plaintext confirmation step provides an extra layer of defense, at the cost of just one additional hash value. This is similar to what done in the Classic McEliece submission [7]. Moreover, the use of *implicit rejection* and a "quiet" KEM (i.e. such that the output is always a session key) further simplifies the API, and is an incentive to design constant-time algorithms, without needing extra machinery or stronger assumptions, as explained in Sects. 14 and 15 of [5].

On the other hand, using Niederreiter has a negative impact on the overall performance of the scheme. The cost of the first step of decoding, detailed above, is comparable to that of reconstructing H' (and computing the syndrome) in the original DAGS, but there is an additional cost in the multiplication by A. Moreover, inverting the matrix C in the second step is expensive, and would slow down decapsulation considerably. In alternative, one could delegate some computation time to the key generation algorithm, and store C^{-1} as private key; this would preserve the efficiency of the decapsulation but noticeably increase the size of the private key. Either way, there is a clear a tradeoff at hand, sacrificing performance and efficiency in favor of a simpler description and tighter security. It therefore falls to the user's discretion whether original DAGS or SimpleDAGS is the best variant to be employed for the purpose.

4   Improved Resistance

It is natural to think that introducing additional algebraic structure like QD in a scheme based on algebraic codes (such as Goppa or GS) can give an adversary more power to perform a structural attack. This is the case of the well-known FOPT attack [10], and successive variants [9,11], which exploit this algebraic structure to solve a multivariate system of equations and reconstruct an alternant matrix which is equivalent to the private key. A detailed analysis of such attacks, and countermeasures, is given in the original DAGS paper. Recently, Barelli and Couvreur presented a structural attack aimed precisely at DAGS [4], which is very successful against the original parameters. We discuss it here.

4.1   The Barelli-Couvreur Attack

The attack makes use of a novel construction called *Norm-Trace Codes*. As the name suggests, these codes are the result of the application of both the Trace and the Norm operation to a certain support vector, and they are alternant codes. In particular, they are subfield subcodes of Reed-Solomon codes. The construction of these codes is given explicitly only for the specific case $m = 2$ (as is the case in all DAGS parameters), i.e. the support vector has components in $\mathbb{F}_{q^2}$, in which case the norm-trace code is defined as

$$\mathcal{NT}(\mathbf{x}) = \langle 1, Tr(\mathbf{x}), Tr(\alpha\mathbf{x}), N(\mathbf{x}) \rangle$$

where α is an element of trace 1.

The main idea is that there exists a specific norm-trace code that is the *conductor* of the secret subcode into the public code. By "conductor" the authors refer to the largest code for which the Schur product (i.e. the component-wise product of all codewords, denoted by $\star$) is entirely contained in the target, i.e.

$$Cond(\mathcal{D}, \mathcal{C}) = \{\mathbf{u} \in \mathbb{F}_q^n : \mathbf{u} \star \mathcal{D} \subseteq \mathcal{C}\}$$

The authors present two strategies to determine the secret subcode. The first strategy is essentially an exhaustive search over all the codes of the proper co-dimension. This co-dimension is given by $2q/s$, since s is the size of the permutation group of the code, which is non-trivial in our case due to the code being quasi-dyadic. While such a brute force in principle would be too expensive, the authors present a few refinements that make it feasible, which include an observation on the code rate of the codes in use, and the use of shortened codes. The second strategy, instead, consists of solving a bilinear system, which is obtained using the parity-check matrix of the public code and treating as unknowns the elements of a generator matrix for the secret code (as well as the support vector $\mathbf{x}$). This system is solved using Gröbner bases techniques, and benefits from a reduction in the number of variables similar to the one performed in FOPT, as well as the refinements mentioned above (shortened codes).

In both cases, it is easy to deduce that the two parameters q and s are crucial in determining the cost of running this step of the attack, which dominates the overall cost. In fact, the authors are able to provide an accurate complexity analysis for the first strategy which confirms this intuition. The average number of iterations of the brute force search is given by q^{2c}, where c is exactly the co-dimension described above, i.e. $c = 2q/s$. In addition, it is shown that the cost of computing Schur products is $2n^3$ operations in the base field. Thus, the overall cost of the recovery step is $2n^3 q^{4q/s}$ operations in $\mathbb{F}_q$. The authors then argue that wrapping up the attack has negligible cost, and that q-ary operations can be done in constant time (using tables) when q is not too big. All this leads to a complexity which is below the desired security level for all of the DAGS parameters that had been proposed at the time of submission. We report these numbers below (Table 2).

Table 2. Early DAGS parameters.

Security level	q	m	n	k	s	t	w	BC 1	BC 2
1	2^5	2	832	416	2^4	13	104	2^{70}	2^{44}
3	2^6	2	1216	512	2^5	11	176	2^{80}	2^{44}
5	2^6	2	2112	704	2^6	11	352	2^{55}	2^{33}

In the last two columns we report, respectively, the complexity of the attack when running it with the first approach (exhaustive search) and the cycle count for the execution of the attack with the second approach (Gröbner bases). The latter was reported in a follow-up paper by Bardet, Bertin, Couvreur and Otmani [3]. As it is possible to observe, the attack complexity is especially low for the last set of parameters since the dyadic order s was chosen to be 2^6, and this is probably too much to provide security against this attack. Still, we point out that, at the time this parameters were proposed, there was no indication this was the case, since this attack is using an entirely new technique, and it is unrelated to the FOPT and folding attacks that we just described.

While the attack performs very well against the original DAGS parameter sets, it is overall not as critical as it appears. In fact, it is shown in Sect. 5.3 of [1] how this can be defeated even by modifying a single parameter, namely the size of the base field q. This is the case for DAGS_3, where changing this value from 2^6 to 2^8 is enough to push the attack complexity beyond the claimed security level. Updated parameters were introduced in [1], and we report them below (Table 4).

Table 3. Updated DAGS parameters.

Security level	q	m	n	k	s	t	w	BC 1
1	2^6	2	832	416	2^4	13	104	$\approx 2^{128}$
3	2^8	2	1216	512	2^5	11	176	$\approx 2^{288}$
5	2^8	2	1600	896	2^5	11	176	$\approx 2^{289}$

Table 4. Memory requirements (bytes).

Parameter set	Public key	Private key	Ciphertext
DAGS_1	8112	2496	656
DAGS_3	11264	4864	1248
DAGS_5	19712	6400	1632

Note that, for DAGS_5, the dyadic order s needed to be amended too, and the rest of the code parameters modified accordingly to respect the requirements on code length, dimension etc. The case of DAGS_1 is a little peculiar. In fact, the theoretical complexity of the first attack approach can be made large enough by simply switching from $q = 2^5$ to $q = 2^6$, similarly to what was done for DAGS_3. With this in mind, and for the sake of simplicity, [1] featured this choice of parameters for DAGS_1, as reported in Table 3. However, thanks to the detailed analysis appeared in [3], it is now possible to see that these parameters are particularly vulnerable to the second attack approach. In what follows, we will briefly explain the reason for this, and present a new choice of parameters for DAGS_1.

As noted before, the success of the attack strongly depends on the dimension of the invariant code $\mathcal{D}$, which is given by $k_0 - c$, where $k_0 = k/s$ is the number of row blocks and $c = 2q/s$ was defined above. For the parameters in question, we have $k_0 = 26$ and $c = 8$ and therefore this dimension is 18. This leads to an imbalance in the ratio of number of equations to number of variables. The former are given by $(k_0 - c)(n_0 - k_0 - 1)$, where $n_0 = n/s$ is the number of column blocks, while the latter consists of the $(k_0 - c)c$ variables of the $\mathbf{U}$ type and the $n_0 - k_0 + c + \log s - 3$ variables of the $\mathbf{V}$ type that define the bilinear system. Therefore we obtain 450 equations in 179 total variables, and this ratio is about 2.5. The authors then show how the system can be solved by specializing the $\mathbf{U}$ variables to obtain linear equations, for a total cost of approximately 2^{111} operations, which is below the claimed security level. Actually, this cost can be further reduced following a hybrid approach that combines exhaustive search and Gröbner bases, to a total of 2^{83}.

The crucial point is that a ratio of 2.5 is quite high, and this is what makes the attack feasible. In contrast, the updated DAGS_5 parameters produce a ratio of 1.1 which is too low (the system has too many variables) while the situation for DAGS_3 is even more extreme, since in this case $c = k_0$ and therefore $\mathcal{D}$ does not even exist. In this case, the authors suggest to use the dual code instead, therefore replacing k_0 with $n_0 - k_0$ in all the above formulas. In principle, this makes the attack applicable, but the parameters yield a ratio of 0.7 which is again too low to be of any use. We insist on this crucial point to select our next choice of parameters for DAGS_1 (where "N.A." stands for "not applicable") (Table 6).

Table 5. New DAGS parameters.

Security level	q	m	n	k	s	t	w	BC 1	BC 2
1	2^8	2	704	352	2^4	11	88	$\approx 2^{542}$	N.A.
3	2^8	2	1216	512	2^5	11	176	$\approx 2^{288}$	N.A.
5	2^8	2	1600	896	2^5	11	176	$\approx 2^{289}$	N.A

Table 6. New memory requirements (bytes).

Parameter set	Public key	Private key	Ciphertext
DAGS_1	7744	2816	736
DAGS_3	11264	4864	1248
DAGS_5	19712	6400	1632

Note that we have only changed the parameters for DAGS_1, but we have chosen to report the other two sets too, in order to provide a complete view. With this new choice, we have $k_0 = 22$ and $c = 32$ and therefore $\mathcal{D}$ does not exist; in fact, not even its dual exists, since in this case $k_0 = n_0 - k_0$. This completely defeats the second attack approach, while the first approach would produce a ridiculously large complexity ($\approx 2^{542}$, see above), and we therefore feel comfortable claiming that DAGS_1 is now safe against all known attacks.

In the end, we can add the Barelli-Couvreur attack(s) to the amount of constraints on the selection of parameters, and we are very thankful to the authors of [4] and [3] for the detailed and careful analysis of the attack techniques. We will provide a complete overview of such constraints in the next section, where we will also detail an entirely new take on the subject.

4.2   Binary DAGS

Parameters in schemes based on QD-GS codes are a carefully balanced machine, needing to satisfy many constraints. First of all, we would like the code dimension $k = n - mst$ to be approximately $n/2$, since having a code rate close to $1/2$ is an optimal choice in many aspects (for instance, against ISD). In second place, in order to stay clear of the Barelli-Couvreur attack, the value q has to be sufficiently large, while the dyadic order s cannot be too big, as we just explained. On the other hand, s should still be large enough to obtain a significant reduction in key size. Balancing these two parameters can be quite hard since both q and s are required to be powers of 2. Meanwhile, the values of the extension degree m and the number of blocks t need to be, jointly, sufficiently large to reach the threshold of $mt > 21$, which is necessary to avoid the FOPT attack. Of course, m, s and t cannot all be large at the same time otherwise the code dimension k would become trivial. Moreover, it is possible to observe that the best outcome is obtained when m and t are of opposite magnitude (one big and one small) rather than both of "medium" size. Now, since s and t are also what determines the number of correctable errors (which is $st/2$), the value of t cannot be too small either, while a small m is helpful to avoid having to work on very large extension fields. Note that q^m still needs to be at least as big as the code length n (since the support elements are required to be distinct). After all these considerations, the result is that, in previous literature [1, 6, 19], the choice of parameters was oriented towards selecting a large base field q and the smallest possible value for the extension, i.e. $m = 2$, with s ranging from 2^4 to 2^6, and t chosen accordingly. We now investigate the consequences of choosing parameters in the opposite way.

Choosing large m and small t allows q to be reduced to the minimum, and more precisely q could be even 2 itself, meaning binary codes are obtained. Binary codes were already considered in the original QD Goppa proposal by Misoczki and Barreto [16], where they ended up being the only safe choice. The reason for this is that larger base fields mean m can be chosen smaller (and in fact, must, in order to avoid working on prohibitively large extension fields). This in turn means FOPT is very effective (remember that there is no parameter t for Goppa codes), so in order to guarantee security one had to choose m as big as possible (at least 16) and consequently $q = 2$. Now in our case, if t is small, s must be bigger (for error-correction purposes), and this pushes n and k up accordingly. We present below our binary parameters (Table 7) and corresponding memory requirements (Table 8).

Table 7. Binary DAGS parameters.

Security level	q	m	n	k	s	t	w	BC
1	2	13	6400	3072	2^7	2	128	N.A.
3	2	14	11520	4352	2^8	2	256	N.A.
5	2	14	14080	6912	2^8	2	256	N.A

Table 8. Memory requirements for binary DAGS (bytes).

Parameter set	Public key	Private key	Ciphertext
DAGS_1	9984	20800	832
DAGS_3	15232	40320	1472
DAGS_5	24192	49280	1792

The parameters are chosen to stay well clear of all the known algebraic attacks. In particular, using binary parameters should entirely prevent the latest attack by Barelli and Couvreur. In this case, in fact, we have $m >> 2$, and it is not yet clear whether the attack is applicable in the first place. However, even if this was the case, the complexity of the attack, which currently depends on the quantity q/s, would depend instead on mq^{m-1}/s. It is obvious that, with our choice of parameters, the attack would be completely infeasible in practice.

Note that, in order to be able to select binary parameters, it is necessary to choose longer codes (as explained above), which end up in slightly larger public keys: these are about 1.3 times those of the original (non-binary) DAGS. The private keys are also considerably larger. On the other hand, the binary base field should bring clear advantages in term of arithmetic, and result in a much more efficient implementation. All things considered, this variant should be seen as yet another tradeoff, in this case sacrificing key size in favor of increased security and efficient implementation.

5   Revised Implementation Results

In this section we present the results obtained in our revised implementation. Our efforts focused on several aspects of the code, with the ultimate goal of providing faster algorithms, but which are also clearer and more accessible. With this in mind, one of the main aspects that was modified is field multiplication.

We removed the table-based multiplication to prevent an easy avenue for side-channel (cache) attacks: this is now vectorized by the compiler, which also allows for a considerable speedup. Moreover, we were able to obtain a considerable gain during key generation by exploiting the diagonal form of the Cauchy matrix. Finally, we "cleaned up" and polished our C code, to ensure it is easier to understand for external auditors. Below, we report timings obtained for our revised implementation (Table 10), as well as the measurements previously obtained for the reference code (Table 9), for ease of comparison. We remark that all these numbers refer to the latest DAGS parameters (i.e. those presented in Table 5); an implementation of Binary DAGS is currently underway. The timings were acquired running the code 100 times and taking the average. We used CLANG compiler version 8.0.0 and the compilation flags -O3 -g3 -Wall -march=native -mtune=native -fomit-frame-pointer -ffast-math. The processor was an Intel(R) Core(TM) i5-5300U CPU @ 2.30 GHz.

Table 9. Timings for reference code.

Algorithm	Cycles		
	DAGS_1	DAGS_3	DAGS_5
Key generation	2,540,311,986	4,320,206,006	7,371,897,084
Encapsulation	12,108,373	26,048,972	96,929,832
Decapsulation	215,710,551	463,849,016	1,150,831,538

Table 10. Timings for revised implementation.

Algorithm	Cycles		
	DAGS_1	DAGS_3	DAGS_5
Key generation	408,342,881	1,560,879,328	2,061,117,168
Encapsulation	5,061,697	14,405,500	35,655,468
Decapsulation	192,083,862	392,435,142	388,316,593

As it is possible to observe, the performance is much faster than the previously reported numbers, despite the increase in parameters (and especially field size). Furthermore, we are planning to obtain even faster timings by incorporating techniques from [2]. These include a dedicated version of the Karatsuba multiplication algorithm (as detailed in [2]), as well as an application of LUP inversion to compute the systematic form of the public key in an efficient manner. Such an implementation is currently underway and results will appear in future work.

6   Conclusion

DAGS was one of the two NIST proposals based on structured algebraic codes, and the only one using Quasi-Dyadic codes (the other, BIG QUAKE, is based on Quasi-Cyclic codes). In fact, DAGS is also the only proposal exploiting the class of Generalized Srivastava codes. As such, we believe DAGS is already an interesting candidate. At the present time, neither of these two proposals was selected to move forward to the second round of the standardization competition. Indeed, BIG QUAKE seemed to privilege security over performance, at the cost of selecting very conservative parameters, which gave way to large keys and ended up pushing the scheme into the same category of "conservative" schemes such as Classic McEliece (where a comparison favored the latter). The approach for DAGS was the opposite: we chose "aggressive" parameters, with the goal of reaching really interesting data sizes. In practice, this meant using $m = 2$ and this led to the Barelli-Couvreur attack. As a consequence, security concerns where raised and this led to the decision of not selecting DAGS for the second round.

In this paper, we investigated several aspects of DAGS, one of which is precisely its security. Exploiting the analysis given in [3], we have shown that two of the updated parameter sets, namely DAGS_3 and DAGS_5, are already beyond the scope of both attack approaches. The third set, that of DAGS_1, was unfortunate as the imbalance between number of equations and number of variables provided a way to instantiate the Gröbner basis technique effectively. Now that an analysis is available, it was easy to select a new parameter set for DAGS_1, which we trust will be the definitive one. We have then provided updated implementation figures, which take into account a variety of ideas for speeding up the code, such as vectorization, and dedicated techniques for Quasi-Dyadic codes. Moreover, we presented two variants offering some tradeoffs. The first is what we call "SimpleDAGS", and it is essentially a conversion of the original protocol to the Niederreiter framework. This allows for a cleaner protocol and a simpler security analysis (as in [5]), at the cost of increased data requirements. The second is a new set of binary parameters, which provides an advantage in terms of security against known structural attacks, again at the cost of a slight increase in data size. While an implementation is still underway for this set of parameters, we expect them to provide much faster times, in line with similar schemes such as Classic McEliece.

References

1. Banegas, G., et al.: DAGS: key encapsulation using dyadic GS codes. J. Math. Cryptol. **12**, 221–239 (2018)
2. Banegas, G., Barreto, P.S.L.M., Persichetti, E., Santini, P.: Designing efficient dyadic operations for cryptographic applications. IACR Cryptology ePrint Archive 2018, p. 650 (2018)
3. Bardet, M., Bertin, M., Couvreur, A., Otmani, A.: Practical algebraic attack on DAGS. To appear

4. Barelli, É., Couvreur, A.: An efficient structural attack on NIST submission DAGS. In: Peyrin, T., Galbraith, S. (eds.) ASIACRYPT 2018. LNCS, vol. 11272, pp. 93–118. Springer, Cham (2018). https://doi.org/10.1007/978-3-030-03326-2_4
5. Bernstein, D.J., Persichetti, E.: Towards KEM unification. IACR Cryptology ePrint Archive 2018, p. 526 (2018)
6. Cayrel, P.-L., Hoffmann, G., Persichetti, E.: Efficient Implementation of a CCA2-Secure Variant of McEliece Using Generalized Srivastava Codes. In: Fischlin, M., Buchmann, J., Manulis, M. (eds.) PKC 2012. LNCS, vol. 7293, pp. 138–155. Springer, Heidelberg (2012). https://doi.org/10.1007/978-3-642-30057-8_9
7. https://classic.mceliece.org/
8. http://www.dags-project.org
9. Faugere, J.-C., Otmani, A., Perret, L., De Portzamparc, F., Tillich, J.-P.: Structural cryptanalysis of McEliece schemes with compact keys. DCC **79**(1), 87–112 (2016)
10. Faugère, J.-C., Otmani, A., Perret, L., Tillich, J.-P.: Algebraic cryptanalysis of McEliece variants with compact keys. In: Gilbert, H. (ed.) EUROCRYPT 2010. LNCS, vol. 6110, pp. 279–298. Springer, Heidelberg (2010). https://doi.org/10.1007/978-3-642-13190-5_14
11. Faugère, J.-C., Otmani, A., Perret, L., Tillich, J.-P:. Algebraic cryptanalysis of McEliece variants with compact keys - towards a complexity analysis. In: Proceedings of the 2nd International Conference on Symbolic Computation and Cryptography, SCC 2010, pp. 45–55. RHUL, June 2010
12. Grover, L.K.: A fast quantum mechanical algorithm for database search. In: Proceedings of the 28th Annual ACM Symposium on the Theory of Computing (STOC), pp. 212–219, May 1996
13. Hofheinz, D., Hövelmanns, K., Kiltz, E.: A modular analysis of the Fujisaki-Okamoto transformation. In: Kalai, Y., Reyzin, L. (eds.) TCC 2017. LNCS, vol. 10677, pp. 341–371. Springer, Cham (2017). https://doi.org/10.1007/978-3-319-70500-2_12
14. https://keccak.team/kangarootwelve.html
15. MacWilliams, F.J., Sloane, N.J.A.: The Theory of Error-Correcting Codes. Elsevier, Amsterdam (1977). North-Holland Mathematical Library
16. Misoczki, R., Barreto, P.S.L.M.: Compact McEliece keys from Goppa codes. In: Jacobson, M.J., Rijmen, V., Safavi-Naini, R. (eds.) SAC 2009. LNCS, vol. 5867, pp. 376–392. Springer, Heidelberg (2009). https://doi.org/10.1007/978-3-642-05445-7_24
17. https://csrc.nist.gov/projects/post-quantum-cryptography/post-quantum-cryptography-standardization
18. Nojima, R., Imai, H., Kobara, K., Morozov, K.: Semantic security for the McEliece cryptosystem without random oracles. Des. Code. Cryptogr. **49**(1–3), 289–305 (2008)
19. Persichetti, E.: Compact McEliece keys based on quasi-dyadic Srivastava codes. J. Math. Cryptol. **6**(2), 149–169 (2012)
20. Sarwate, D.: On the complexity of decoding Goppa codes. IEEE Trans. Inf. Theory **23**(4), 515–516 (1977)
21. Shor, P.W.: Polynomial-time algorithms for prime factorization and discrete logarithms on a quantum computer. SIAM J. Comput. **26**(5), 1484–1509 (1997)

Practical Algebraic Attack on DAGS

Magali Bardet[1(✉)], Manon Bertin[1], Alain Couvreur[2], and Ayoub Otmani[1]

[1] LITIS, University of Rouen Normandie Avenue de l'université,
76801 Saint-Étienne-du-Rouvray, France
{`magali.bardet,manon.bertin8,Ayoub.Otmani`}`@univ-rouen.fr`
[2] INRIA & LIX, CNRS UMR 7161 École polytechnique,
91128 Palaiseau Cedex, France
`alain.couvreur@lix.polytechnique.fr`

Abstract. DAGS scheme is a key encapsulation mechanism (KEM) based on quasi-dyadic alternant codes that was submitted to NIST standardization process for a quantum resistant public key algorithm. Recently an algebraic attack was devised by Barelli and Couvreur (Asiacrypt 2018) that efficiently recovers the private key. It shows that DAGS can be totally cryptanalysed by solving a system of bilinear polynomial equations. However, some sets of DAGS parameters were not broken in practice. In this paper we improve the algebraic attack by showing that the original approach was not optimal in terms of the ratio of the number of equations to the number of variables. Contrary to the common belief that reducing at any cost the number of variables in a polynomial system is always beneficial, we actually observed that, provided that the ratio is increased and up to a threshold, the solving can be heavily improved by adding variables to the polynomial system. This enables us to recover the private keys in a few seconds. Furthermore, our experimentations also show that the maximum degree reached during the computation of the Gröbner basis is an important parameter that explains the efficiency of the attack. Finally, the authors of DAGS updated the parameters to take into account the algebraic cryptanalysis of Barelli and Couvreur. In the present article, we propose a hybrid approach that performs an exhaustive search on some variables and computes a Gröbner basis on the polynomial system involving the remaining variables. We then show that the updated set of parameters corresponding to 128-bit security can be broken with 2^{83} operations.

Keywords: Quantum safe cryptography · McEliece cryptosystem · Algebraic cryptanalysis · Dyadic alternant code

1 Introduction

The design of a quantum-safe public key encryption scheme is becoming an important issue with the recent process initiated by NIST to standardize one or more quantum-resistant public-key cryptographic algorithms. One of the oldest

© Springer Nature Switzerland AG 2019
M. Baldi et al. (Eds.): CBC 2019, LNCS 11666, pp. 86–101, 2019.
https://doi.org/10.1007/978-3-030-25922-8_5

cryptosystem that is not affected by the apparition of a large-scale quantum computer is the McEliece public key encryption scheme [26]. It is a code-based cryptosystem that uses the family of binary Goppa codes. The main advantage of this cryptosystem is its very fast encryption/decryption functions, and the fact that up to the present, nobody has succeeded to cryptanalyse it.

But in the eyes of those who are concerned with applications requiring very compact schemes, these positive aspects of the McEliece cryptosystem may not make up for its large keys. For instance the classic McEliece [10] submitted to NIST uses at least 1MB public keys for 256 bits of security. A well-known approach for getting smaller keys consists in replacing binary Goppa codes by even more structured linear codes. A famous method started in [23] and further developed in [3,6,9,27] relies on codes displaying symmetries like cyclicity and dyadicity while having very efficient decoding algorithms. Unlike the McEliece cryptosystem which currently remains unbroken, the schemes [9,27] are subject to efficient ciphertext-only attacks [20] that recover the secret algebraic structure. The attack developed in [20] formulates the general problem of recovering the algebraic structure of an alternant code as solving a system of polynomial equations. But it involves very high degree polynomial equations with too many variables. Finding solutions to this kind of algebraic system is currently out of reach of the best known algorithms. However this approach turns out to be extremely fruitful when dealing with polynomial systems that come from the quasi-cyclic [9] and quasi-dyadic [27] cryptosystems because the symmetries permit to reduce to a manageably small number of variables.

The apparition of the algebraic attack in [20] generated a series of new algebraic attacks [17,19,21] but since the original McEliece cryptosystem does not seem to be affected by this approach, it still raises the question of whether it represents a real threat.

Recently a new algebraic attack [8] was mounted against DAGS [3] scheme. DAGS is a key encapsulation mechanism (KEM) based on quasi-dyadic alternant codes defined over quadratic extension. It was submitted to the standardization process launched by NIST. The attack relies on the component-wise product of codes in order to build a system of bilinear multivariate equations. The use of the component-wise product of codes in cryptography is not new. It first appeared in [29] and has proved in several occasions [14,24,25,28] to be a powerful cryptanalytic tool against algebraic codes like Generalised Reed-Solomon codes. It even enabled to mount for the first time a polynomial-time attack in [15,16] against a special family of non-binary Goppa codes [11] displaying no symmetries.

Our Contribution. In this paper we improve the algebraic attack of [8] by showing that the original approach was not optimal in terms of the ratio of the number of equations to the number of variables. Contrary to the common belief that reducing at any cost the number of variables in a polynomial system is always beneficial, we actually observed that, provided that the ratio is increased and up to a threshold, the solving can be heavily improved by adding variables to the polynomial system. This enables us to recover the private keys in a few seconds. In Table 1 we report the average running times of our attack and that of [8] performed on the same machine. For DAGS-1 and DAGS-5, the linear algebra part of the attack is the dominant cost.

Furthermore, our experimentation show that the maximum degree reached during the computation of the Gröbner basis is an important parameter that explains the efficiency of the attack. We observed that the maximum degree never exceeds 4.

Subsequently to the attack [8], the authors of DAGS updated the parameters. We propose a hybrid approach that performs an exhaustive search on some variables and computes a Gröbner basis on the polynomial system involving the remaining variables. We then show that one set of parameters does not have the claimed level of security. Indeed the parameters corresponding to 128-bit security can be broken with 2^{83} operations.

Table 1. Running times of the algebraic attack to break DAGS scheme. The computations are performed with Magma V2.23-1 on an Intel Xeon processor clocked at 2.60 GHz with 128 Gb. We reproduced the computations from [8] on our machine. The columns "Gröbner" correspond to the Gröbner basis computation part, the columns "Linear algebra" to the linear algebra steps of the attack.

Parameters	Security	[8]			The present article		
		Gröbner	Linear algebra	Total	Gröbner	Linear algebra	Total
DAGS-1	128	552 s	8 s	560 s	3.6 s	6.4 s	10 s
DAGS-3	192	–	–	–	70 s	16 s	86 s
DAGS-5	256	6 s	20 s	26 s	0.5 s	15.5 s	16 s

Organization of the Paper. Section 2 introduces the useful notation and important notions to describe the DAGS scheme. Section 3 recalls the important properties about the component-wise product of GRS and alternant codes. In Sect. 4 we describe the algebraic attack, and in Sect. 5 the practical running times we obtained with our experimentations. Lastly, Sect. 6 explains the hybrid approach.

2 Preliminaries

Notation. $\mathbb{F}_q$ is the field with q elements. In this paper, q is a power of 2. For any m and n in $\mathbb{Z}$, $[\![m, n]\!]$ is the set of integers i such that $m \leqslant i \leqslant n$. The cardinality of set A is $|A|$. Vectors and matrices are denoted by boldface letters as $\boldsymbol{a} = (a_1, \ldots, a_n)$ and $\boldsymbol{A} = (a_{i,j})$. The *Symmetric* group on n letters is denoted by $\mathfrak{S}_n$ and for any $\boldsymbol{v} = (v_1, \ldots, v_n)$ and σ in $\mathfrak{S}_n$ we define $\boldsymbol{v}^\sigma \triangleq (v_{\sigma(1)}, \ldots, v_{\sigma(n)})$. The *identity* matrix of size n is written as $\boldsymbol{I}_n$. The transpose of a vector $\boldsymbol{a}$ and a matrix $\boldsymbol{A}$ is denoted by $\boldsymbol{a}^T$ and $\boldsymbol{A}^T$. The i-th row of a matrix $\boldsymbol{A} = (a_{i,j})$ is $\boldsymbol{A}[i]$. We recall that the *Kronecker product* $\boldsymbol{a} \otimes \boldsymbol{b}$ of two vectors $\boldsymbol{a} = (a_1, \ldots, a_n)$ and

b is equal to $(a_1 b, \ldots, a_n b)$. In particular, we denote by $\mathbb{1}_n$ the all–one vector $\mathbb{1}_n \triangleq (1, \ldots, 1) \in \mathbb{F}_q^n$ and we have $\mathbb{1}_n \otimes \boldsymbol{a} = (\boldsymbol{a}, \ldots, \boldsymbol{a})$ and

$$\boldsymbol{a} \otimes \mathbb{1}_n = (a_1, \ldots, a_1, a_2, \ldots, a_2, \ldots, a_n, \ldots, a_n).$$

Any k-dimensional vector subspace $\mathscr{C}$ of $\mathbb{F}^n$ where $\mathbb{F}$ is field is called a *linear code* $\mathscr{C}$ *of length* n and *dimension* $k < n$ over $\mathbb{F}$. A matrix whose rows form a basis of $\mathscr{C}$ is called a *generator matrix*. The *orthogonal* or *dual* of $\mathscr{C} \subset \mathbb{F}^n$ is the linear space $\mathscr{C}^\perp$ containing all vectors $\boldsymbol{z}$ from $\mathbb{F}^n$ such that for all $\boldsymbol{c} \in \mathscr{C}$, we have $\langle \boldsymbol{c}, \boldsymbol{z} \rangle \triangleq \sum_{i=1}^n c_i z_i = 0$. We always have $\dim \mathscr{C}^\perp = n - \dim \mathscr{C}$, and any generator matrix of $\mathscr{C}^\perp$ is called a *parity check* matrix of U. The *punctured* code $\mathcal{P}_I(\mathscr{C})$ of $\mathscr{C}$ over a set $I \subset [\![1, n]\!]$ is defined as

$$\mathcal{P}_I(\mathscr{C}) \triangleq \left\{ \boldsymbol{u} \in \mathbb{F}^{n-|I|} \mid \exists \boldsymbol{c} \in \mathscr{C}, \ \boldsymbol{u} = (c_i)_{i \in [\![1,n]\!] \setminus I} \right\}.$$

The *shortened* $\mathcal{S}_I(\mathscr{C})$ code of $\mathscr{C}$ over a set $I \subset [\![1, n]\!]$ is then defined as

$$\mathcal{S}_I(\mathscr{C}) \triangleq \mathcal{P}_I \left(\left\{ \boldsymbol{c} \in \mathscr{C} \mid \forall i \in I, \ c_i = 0 \right\} \right).$$

We extend naturally the notation $\mathcal{P}_I(\cdot)$ to vectors and matrices.

Algebraic codes. A *generalized Reed-Solomon code* $\mathsf{GRS}_t(\boldsymbol{x}, \boldsymbol{y})$ of dimension t and length n where $\boldsymbol{x}$ is an n-tuple of distinct elements from a finite field $\mathbb{F}$ and $\boldsymbol{y}$ is an n-tuple of non-zero elements from $\mathbb{F}$ is the linear code defined by

$$\mathsf{GRS}_t(\boldsymbol{x}, \boldsymbol{y}) \triangleq \left\{ (y_1 f(x_1), \ldots, y_n f(x_n)) \mid f \in \mathbb{F}_{<t}[z] \right\} \tag{1}$$

where $\mathbb{F}_{<t}[z]$ is the set of univariate polynomials f with coefficients in $\mathbb{F}$ such that $\deg f < t$. The dimension of $\mathsf{GRS}_t(\boldsymbol{x}, \boldsymbol{y})$ is clearly t. By convention $\mathsf{GRS}_t(\boldsymbol{x}, \mathbb{1}_n)$ where $\mathbb{1}_n$ is the all-one vector of length n is simply a Reed-Solomon code denoted by $\mathsf{RS}_t(\boldsymbol{x})$.

The code $\mathsf{GRS}_t{}^\perp(\boldsymbol{x}, \boldsymbol{y})$ is equal to $\mathsf{GRS}_{n-t}(\boldsymbol{x}, \boldsymbol{y}^\perp)$ where $\boldsymbol{y}^\perp = (y_1^\perp, \ldots, y_n^\perp)$ is the n-tuple such that for all j in $[\![1, n]\!]$ it holds that

$$\left(y_j^\perp \right)^{-1} = y_j \prod_{\ell=1, \ell \neq j}^n (x_\ell - x_j). \tag{2}$$

An *alternant code* $\mathscr{A}_t(\boldsymbol{x}, \boldsymbol{y})$ of degree $t \geqslant 1$ over a field $\mathbb{K} \subsetneq \mathbb{F}$ and length n where x is an n-tuple of distinct elements from $\mathbb{F}^n$ and $\boldsymbol{y}$ is an n-tuple of non-zero elements from $\mathbb{F}^n$ is the linear code

$$\mathscr{A}_t(\boldsymbol{x}, \boldsymbol{y}) \triangleq \mathsf{GRS}_{n-t}(\boldsymbol{x}, \boldsymbol{y}^\perp) \cap \mathbb{K}^n = \mathsf{GRS}_t^\perp(\boldsymbol{x}, \boldsymbol{y}) \cap \mathbb{K}^n. \tag{3}$$

The dimension of an alternant code satisfies the bound $\dim \mathscr{A}_t(\boldsymbol{x}, \boldsymbol{y}) \geqslant n - mt$ where m is the degree of the field extension of $\mathbb{F}/\mathbb{K}$.

Remark 1. Note that one has always the inclusions $\mathsf{GRS}_r(\boldsymbol{x}, \boldsymbol{y}) \subseteq \mathsf{GRS}_{t+r}(\boldsymbol{x}, \boldsymbol{y})$ and $\mathscr{A}_{t+r}(\boldsymbol{x}, \boldsymbol{y}) \subseteq \mathscr{A}_r(\boldsymbol{x}, \boldsymbol{y})$ for any $r \geqslant 1$ and $t \geqslant 1$.

Proposition 1. *Let* $\mathsf{GRS}_k(\boldsymbol{x}, \boldsymbol{y})$ *be a generalized Reed-Solomon code of dimension* k *where* $\boldsymbol{x}$ *is an* n*-tuple of distinct elements from* $\mathbb{F}_{q^m}$ *and* $\boldsymbol{y}$ *is an* n*-tuple of non-zero elements from* $\mathbb{F}_{q^m}$. *For any affine map* $\zeta : \mathbb{F}_{q^m} \to \mathbb{F}_{q^m}$ *defined as* $\zeta(z) \triangleq az + b$ *where* a *in* $\mathbb{F}_{q^m} \setminus \{0\}$ *and* b *in* $\mathbb{F}_{q^m}$ *it holds that*

$$\mathsf{GRS}_k\big(\zeta(\boldsymbol{x}), \boldsymbol{y}\big) = \mathsf{GRS}_k(\boldsymbol{x}, \boldsymbol{y}).$$

Remark 2. A consequence of Proposition 1 is that it is possible to choose arbitrary values for two different coordinates x_i and x_j ($i \neq j$) provided that they are different. For instance we may always assume that $x_1 = 0$ and $x_2 = 1$.

Another very important result from this proposition is that when an affine map $\zeta(z) = az + b$ leaves globally invariant $\boldsymbol{x}$ then it induces a permutation σ of $\mathfrak{S}_n$ thanks to the identification:

$$\forall i \in [\![1, n]\!], \quad x_{\sigma(i)} \triangleq \zeta(x_i).$$

We call σ the permutation *induced* by the affine map ζ. For the ease of notation we shall systematically identify σ and ζ.

Dyadic Codes. A code $\mathscr{C} \subset \mathbb{F}^n$ is *quasi-dyadic of order* 2^γ where γ is a non-negative integer if there exists $\mathbb{G} \subseteq \mathfrak{S}_n$ that is isomorphic to $\mathbb{F}_2^\gamma$ such that

$$\forall(\sigma, c) \in \mathbb{G} \times \mathscr{C}, \quad c^\sigma \in \mathscr{C}.$$

A construction of quasi-dyadic GRS and alternant codes is given in [18]. It considers γ elements $b_1, \ldots, b_\gamma$ in $\mathbb{F}_{q^m}$ that are linearly independent over $\mathbb{F}_2$. The vector space $\oplus_{i=1}^\gamma \mathbb{F}_2 \cdot b_i$ generated over $\mathbb{F}_2$ is then equal to a group $\mathbb{G}$. Next, it takes an n_0-tuple $\boldsymbol{\tau} = (\tau_1 \ldots, \tau_{n_0})$ from $\mathbb{F}_{q^m}^{n_0}$ such that the cosets $\tau_i + \mathbb{G}$ are pairwise disjoint, and finally it picks an n_0-tuple $\boldsymbol{y} = (y_1, \ldots, y_{n_0})$ composed of nonzero elements from $\mathbb{F}_{q^m}$. We now consider $\boldsymbol{z} \triangleq \boldsymbol{y} \otimes \mathbb{1}_{2^\gamma}$ and $\boldsymbol{x} \triangleq \boldsymbol{\tau} \otimes \mathbb{1}_{2^\gamma} + \mathbb{1}_{n_0} \otimes \boldsymbol{g}$ where $\boldsymbol{g} \triangleq (g)_{g \in \mathbb{G}}$. The action of $\mathbb{G}$ can then be described more explicitly: for any b in $\mathbb{G}$ we associate the translation defined for any z in $\mathbb{F}_{q^m}$ by $\sigma_b(z) \triangleq z + b$. It is clear that σ_b leaves globally invariant $\boldsymbol{x}$ because we have $\sigma_b(\mathbb{G}) = b + \mathbb{G} = \mathbb{G}$ and furthermore the following holds

$$\sigma_b(\boldsymbol{x}) = \boldsymbol{x} + b \otimes \mathbb{1}_{2^\gamma n_0} = \boldsymbol{\tau} \otimes \mathbb{1}_{2^\gamma} + \mathbb{1}_{n_0} \otimes \sigma_b(\boldsymbol{g}).$$

Proposition 2 [18]. *Let* $\mathbb{K}$ *be a subfield of* $\mathbb{F}_{q^m}$, n_0, γ, $\mathbb{G}$, $\boldsymbol{y}$ *and* $\boldsymbol{\tau}$ *defined as above and* $n \triangleq 2^\gamma n_0$, $\boldsymbol{g} \triangleq (g)_{g \in \mathbb{G}}$, $\boldsymbol{z} \triangleq \boldsymbol{y} \otimes \mathbb{1}_{2^\gamma}$ *and* $\boldsymbol{x} \triangleq \boldsymbol{\tau} \otimes \mathbb{1}_{2^\gamma} + \mathbb{1}_{n_0} \otimes \boldsymbol{g}$. *The codes* $\mathsf{GRS}_r(\boldsymbol{x}, \boldsymbol{z}) \subset \mathbb{F}_{q^m}^n$ *and* $\mathscr{A}_t(\boldsymbol{x}, \boldsymbol{z}) \subset \mathbb{K}^n$ *are quasi-dyadic of order* 2^γ.

Example 1. Let us take $n_0 = 2$ and $\gamma = 2$ then $\boldsymbol{g} = (0, b_1, b_2, b_1 + b_2)$ and $\boldsymbol{x} = (\tau_1, \tau_1 + b_1, \tau_1 + b_2, \tau_1 + b_1 + b_2) || (\tau_2, \tau_2 + b_1, \tau_2 + b_2, \tau_2 + b_1 + b_2)$. The group $\mathbb{G}$ is then equal to $\{0, b_1, b_2, b_1 + b_2\}$. We have $\sigma_{b_1}(\boldsymbol{g}) = (b_1, 0, b_2 + b_1, b_2)$ and the permutation that corresponds to b_1 is $(12)(34)(56)(78)$ in the canonical cycle notation. □

DAGS Scheme. The public key encryption scheme DAGS [1,2] submitted to the NIST call for post quantum cryptographic proposals is a McEliece-like scheme with a conversion to a KEM. It relies on quasi-dyadic alternant codes $\mathscr{A}_r = \mathsf{GRS}_r^{\perp}(\boldsymbol{x}, \boldsymbol{z}) \cap \mathbb{F}_q^n$ with $q = 2^s$ with $\mathsf{GRS}_r^{\perp}(\boldsymbol{x}, \boldsymbol{z}) \subset \mathbb{F}_{q^2}^n$ $(m = 2)$. The public code $\mathscr{A}_r$ is quasi-dyadic of order 2^{γ} where $\gamma \geqslant 1$. The parameters are chosen such that $r = 2^{\gamma} r_0$ and $n = 2^{\gamma} n_0$, and the dimension is $k = 2^{\gamma} k_0$ with $k_0 \triangleq n_0 - 2 r_0$.

Keeping up with the notation of Proposition 2 the vectors $\boldsymbol{x}$ and $\boldsymbol{z}$ can be written as $\boldsymbol{x} \triangleq \boldsymbol{\tau} \otimes \mathbb{1}_{2^{\gamma}} + \mathbb{1}_{n_0} \otimes \boldsymbol{g}$ and $\boldsymbol{z} \triangleq \boldsymbol{y} \otimes \mathbb{1}_{2^{\gamma}}$ with $\boldsymbol{g} = (g)_{g \in \mathbb{G}}$ where $\mathbb{G} = \oplus_{i=1}^{\gamma} \mathbb{F}_2 b_i$ is the vector space generated over $\mathbb{F}_2$ by γ elements $\boldsymbol{b} = (b_1, \ldots, b_{\gamma})$ that are linearly independent over $\mathbb{F}_2$. The quantities $\boldsymbol{b}$ and $\boldsymbol{y}$ are randomly drawn from $\mathbb{F}_{q^2}^{n_0}$ such that the cosets $\tau_i + \mathbb{G}$ are pairwise disjoint and $\boldsymbol{y}$ is composed of nonzero elements in $\mathbb{F}_{q^2}$.

The public key is then an $(n - k) \times n$ parity check matrix $\boldsymbol{H}_{\mathsf{pub}}$ of $\mathscr{A}_r$. The quantities $(\boldsymbol{b}, \boldsymbol{\tau}, \boldsymbol{y})$ have to be kept secret since they permit to decrypt any ciphertext. Table 2 gathers the parameters of the scheme.

Table 2. DAGS-1, DAGS-3 and DAGS-5 correspond to the initial parameters (v1). When the algebraic attack [8] appeared the authors updated to DAGS-1.1, DAGS-3.1 and DAGS-5.1 (v2).

Name	Security	q	m	2^{γ}	n_0	k_0	r_0
DAGS-1	128	2^5	2	2^4	52	26	13
DAGS-3	192	2^6	2	2^5	38	16	11
DAGS-5	256	2^6	2	2^6	33	11	11
DAGS-1.1	128	2^6	2	2^4	52	26	13
DAGS-3.1	192	2^8	2	2^5	38	16	11
DAGS-5.1	256	2^8	2	2^5	50	28	11

3   Component-Wise Product of Codes

An important property about GRS codes is that whenever $\boldsymbol{a}$ belongs to $\mathsf{GRS}_r(\boldsymbol{x}, \boldsymbol{y})$ and $\boldsymbol{b}$ belongs to $\mathsf{GRS}_t(\boldsymbol{x}, \boldsymbol{z})$, the component wise product $\boldsymbol{a} \star \boldsymbol{b} \triangleq (a_1 b_1, \ldots, a_n b_n)$ belongs to $\mathsf{GRS}_{t+r-1}(\boldsymbol{x}, \boldsymbol{y} \star \boldsymbol{z})$. Furthermore, if one defines the component wise product $\mathscr{A} \star \mathscr{B}$ of two linear codes $\mathscr{A} \subset \mathbb{F}^n$ and $\mathscr{B} \subset \mathbb{F}^n$ as the linear code spanned by all the products $\boldsymbol{a} \star \boldsymbol{b}$ with $\boldsymbol{a}$ in $\mathscr{A}$ and $\boldsymbol{b}$ in $\mathscr{B}$ the inclusion is then an equality

$$\mathsf{GRS}_r(\boldsymbol{x}, \boldsymbol{y}) \star \mathsf{GRS}_t(\boldsymbol{x}, \boldsymbol{z}) = \mathsf{GRS}_{r+t-1}(\boldsymbol{x}, \boldsymbol{y} \star \boldsymbol{z}). \tag{4}$$

In the case of alternant codes over a subfield $\mathbb{K} \subseteq \mathbb{F}$, one only gets in general the inclusion

$$\left(\mathsf{GRS}_{n-r}(\boldsymbol{x}, \boldsymbol{y}^\perp) \cap \mathbb{K}^n\right) \star \left(\mathsf{GRS}_t(\boldsymbol{x}, \mathbb{1}_n) \cap \mathbb{K}^n\right) \subseteq \mathsf{GRS}_{n-r+t-1}(\boldsymbol{x}, \boldsymbol{y}^\perp) \cap \mathbb{K}^n \tag{5}$$

which leads to the following result.

Proposition 3 [8]. *For any integer* $r \geqslant 1$ *and* $t \geqslant 1$ *the alternant codes* $\mathscr{A}_{r+t-1}(\boldsymbol{x}, \boldsymbol{y})$ *and* $\mathscr{A}_r(\boldsymbol{x}, \boldsymbol{y})$ *over* $\mathbb{K} \subsetneq \mathbb{F}$ *where* $\boldsymbol{x}$ *is an* n-*tuple of distinct elements from a finite field* $\mathbb{F}$ *and* $\boldsymbol{y}$ *is an* n-*tuple of non-zero elements from* $\mathbb{F}$ *satisfy the inclusion*

$$\mathscr{A}_{r+t-1}(\boldsymbol{x}, \boldsymbol{y}) \star (\mathsf{RS}_t(\boldsymbol{x}) \cap \mathbb{K}^n) \subseteq \mathscr{A}_r(\boldsymbol{x}, \boldsymbol{y}). \tag{6}$$

The previous result is really interesting when $\mathsf{RS}_t(\boldsymbol{x}) \cap \mathbb{K}^n$ is not the (trivial) code generated by $\mathbb{1}_n$. This happens for instance for $\mathbb{F} = \mathbb{F}_{q^m}$ and $\mathbb{K} = \mathbb{F}_q$ when $t = q^{m-1} + 1$ because $\mathsf{RS}_t(\boldsymbol{x}) \cap \mathbb{F}_q^n$ then always contains at least $\mathbb{1}_n$ and $\mathrm{T}_{\mathbb{F}_{q^m}/\mathbb{F}_q}(\boldsymbol{x}) \triangleq \left(\mathrm{T}_{\mathbb{F}_{q^m}/\mathbb{F}_q}(x_1), \ldots, \mathrm{T}_{\mathbb{F}_{q^m}/\mathbb{F}_q}(x_n)\right)$. Actually one can observe that $\mathrm{T}_{\mathbb{F}_{q^m}/\mathbb{F}_q}(\alpha \boldsymbol{x})$ also always belongs to $\mathsf{RS}_t(\boldsymbol{x}) \cap \mathbb{F}_q^n$ for all α in $\mathbb{F}_{q^m}$. Hence if $\{1, \omega_1, \ldots, \omega_{m-1}\}$ form an $\mathbb{F}_q$-basis of $\mathbb{F}_{q^m}$ then α can be written as $\alpha_0 + \alpha_1 \omega_1 + \cdots + \alpha_{m-1}\omega_{m-1}$ with $\alpha_0, \ldots, \alpha_{m-1}$ in $\mathbb{F}_q$, and consequently with the convention that $\omega_0 \triangleq 1$ one has that

$$\mathrm{T}_{\mathbb{F}_{q^m}/\mathbb{F}_q}(\alpha \boldsymbol{x}) = \sum_{i=0}^{m-1} \alpha_i \mathrm{T}_{\mathbb{F}_{q^m}/\mathbb{F}_q}(\omega_i \boldsymbol{x}).$$

This implies that $\dim_{\mathbb{F}_q} \mathsf{RS}_t(\boldsymbol{x}) \cap \mathbb{F}_q^n \geq m+1$ when $t = q^{m-1}+1$. Another interesting case is when $t = \frac{q^m-1}{q-1}+1$ then $\mathrm{N}_{\mathbb{F}_{q^m}/\mathbb{F}_q}(\boldsymbol{x}) \triangleq \left(\mathrm{N}_{\mathbb{F}_{q^m}/\mathbb{F}_q}(x_1), \ldots, \mathrm{N}_{\mathbb{F}_{q^m}/\mathbb{F}_q}(x_n)\right)$ belongs to $\mathsf{RS}_t(\boldsymbol{x}) \cap \mathbb{F}_q^n$, and one would get that $\dim_{\mathbb{F}_q} \mathsf{RS}_t(\boldsymbol{x}) \cap \mathbb{F}_q^n \geqslant m + 2$.

4   Algebraic Cryptanalysis

We present the ciphertext-only attack of [8] that recovers the private key of DAGS scheme. We refer to Sect. 2 for the notation. The public key is a parity-check matrix $\boldsymbol{H}_{\mathsf{pub}}$ of a quasi-dyadic alternant code $\mathscr{A}_r$. The attack recovers the secret values $\boldsymbol{b} = (b_1, \ldots, b_\gamma)$, $\boldsymbol{\tau} = (\tau_1, \ldots, \tau_{n_0})$ and $\boldsymbol{y} = (y_1, \ldots, y_{n_0})$. The idea is to exploit the fact that

$$\mathscr{A}_{r+t-1}(\boldsymbol{x}, \boldsymbol{y} \otimes \mathbb{1}_{2^\gamma}) \star \left(\mathsf{RS}_t(\boldsymbol{x}) \cap \mathbb{F}_q^n\right) \subseteq \mathscr{A}_r(\boldsymbol{x}, \boldsymbol{y} \otimes \mathbb{1}_{2^\gamma}). \tag{7}$$

where $\boldsymbol{x} = \boldsymbol{\tau} \otimes \mathbb{1}_{2^\gamma} + \mathbb{1}_{n_0} \otimes \boldsymbol{g}$ with $\boldsymbol{g} = (g)_{g \in \mathbb{G}}$ and $\mathbb{G} = \oplus_{i=1}^\gamma \mathbb{F}_2 b_i$. Because the secret vector $\boldsymbol{y}$ is not anymore involved in the definition of $\mathsf{RS}_t(\boldsymbol{x})$ an attacker gains a real advantage if she manages to identify the codewords that are contained in $\mathsf{RS}_t(\boldsymbol{x}) \cap \mathbb{F}_q^n$, especially when $t \geqslant q + 2$ (see [8] for more details). The attack of [8] introduces the *invariant code* $\mathscr{A}_r^{\mathbb{G}}$ *with respect to* $\mathbb{G}$ of $\mathscr{A}_r$ which is defined as

$$\mathscr{A}_r^{\mathbb{G}} \triangleq \left\{ (c_1, \ldots, c_n) \in \mathscr{A}_r \;\middle|\; \forall (i,j) \in [\![0, n_0 - 1]\!] \times [\![1, 2^\gamma]\!], \;\; c_{i2^\gamma + j} = c_{i2^\gamma + 1} \right\}.$$

The dimension of $\mathscr{A}_r^{\mathbb{G}}$ is equal to k_0 (see [7,8]). The cryptanalysis relies then on finding two vector spaces $\mathscr{D}$ and $\mathscr{N}$ such that the constraints given in (8) hold

$$\begin{cases} \mathscr{D} \subsetneq \mathscr{A}_r^{\mathbb{G}}, \\ \dim \mathscr{D} = k_0 - c \;\; \text{where } c \triangleq \frac{mq^{m-1}}{2^\gamma} = \frac{q}{2^{\gamma-1}}, \\ \mathscr{D} \star \mathscr{N} \subseteq \mathscr{A}_r. \end{cases} \tag{8}$$

Let us recall that $\mathscr{A}_r$ and $\mathscr{A}_r^{\mathbb{G}}$ are known, especially it is simple to compute a generator matrix $\boldsymbol{G}_{\mathsf{inv}}$ of $\mathscr{A}_r^{\mathbb{G}}$. Since $\mathscr{D} \subsetneq \mathscr{A}_r^{\mathbb{G}}$ and $\dim \mathscr{D} = k_0 - c$ there exists a $(k_0 - c) \times k_0$ matrix $\boldsymbol{K}$ such that $\boldsymbol{K}\boldsymbol{G}_{\mathsf{inv}}$ generates $\mathscr{D}$. On the other hand $\mathscr{N}$ necessarily satisfies the inclusion[1] $\mathscr{N} \subseteq \left(\mathscr{D} \star \mathscr{A}_r^\perp \right)^\perp$ and consequently $\left(\boldsymbol{K}\boldsymbol{G}_{\mathsf{inv}} \right) \star \boldsymbol{H}_{\mathsf{pub}}$ is a parity check matrix of $\mathscr{N}$. We are now able to state (without proof) an important result justifying the interest of this approach.

Theorem 1 [8]. *Let us assume that* $|\mathbb{G}| \leqslant q$. *Let* $\mathscr{D}$ *be the invariant code of* $\mathscr{A}_{r+q}(\boldsymbol{x}, \boldsymbol{z})$ *and let* $\mathscr{N}$ *be the vector space generated over* $\mathbb{F}_q$ *by* $\mathbb{1}_n$, $\mathrm{T}_{\mathbb{F}_{q^2}/\mathbb{F}_q}(\boldsymbol{x})$ $\mathrm{T}_{\mathbb{F}_{q^2}/\mathbb{F}_q}(\omega \boldsymbol{x})$ *and* $\mathrm{N}_{\mathbb{F}_{q^2}/\mathbb{F}_q}(\boldsymbol{x})$ *where* $\{1, \omega\}$ *is an* $\mathbb{F}_q$-*basis of* $\mathbb{F}_{q^2}$. *Then* $\mathscr{D}$ *and* $\mathscr{N}$ *are solution to* (8).

Remark 3. Considering now the vector space generated by $\mathscr{N}$ over $\mathbb{F}_{q^2}$ one can see that $\boldsymbol{x}$ is also solution to (8) using this simple identity

$$\boldsymbol{x} = (\omega^q - \omega)^{-1} \left(\omega^q \mathrm{T}_{\mathbb{F}_{q^2}/\mathbb{F}_q}(\boldsymbol{x}) - \mathrm{T}_{\mathbb{F}_{q^2}/\mathbb{F}_q}(\omega \boldsymbol{x}) \right).$$

The algebraic attack of [8] recovers $\mathscr{D}$ and $\mathscr{N}$ satisfying (8) by introducing two sets of variables $\boldsymbol{V} = (V_1, \ldots, V_n)$ and $\boldsymbol{K} = (K_{i,j})$ with $i \in [\![1, k_0 - c]\!]$ and $j \in [\![1, k_0]\!]$ that satisfy the multivariate quadratic system

$$\left(\boldsymbol{K}\boldsymbol{G}_{\mathsf{inv}} \right) \star \boldsymbol{H}_{\mathsf{pub}} \cdot \boldsymbol{V}^T = \boldsymbol{0}.$$

The number of variables of this system can be very high which is a hurdle to solving it efficiently in practice. However this algebraic system does not take into account three observations that enable us to significantly reduce the number of variables.

---

[1] It was observed experimentally in [8] that actually the inclusion is most of the time an equality.

- We know by Theorem 1 that $T_{\mathbb{F}_{q^2}/\mathbb{F}_q}(\boldsymbol{x})$ (and $\boldsymbol{x}$) are solution to (8) which means that we may assume that $\boldsymbol{V}$ has a "quasi-dyadic" structure. We define two sets of variables $\boldsymbol{T} = (T_1, \ldots, T_{n_0})$ and $\boldsymbol{B} = (B_1, \ldots, B_\gamma)$ so that we can write $\boldsymbol{V} = \boldsymbol{T} \otimes \mathbb{1}_{2^\gamma} + \mathbb{1}_{n_0} \otimes (\mathbb{F}_2 \cdot \boldsymbol{B})$ where $\mathbb{F}_2 \cdot FB \triangleq \oplus_{i=1}^\gamma \mathbb{F}_2 B_i$ is the vector formed by all the elements in the $\mathbb{F}_2$-vector space generated by $\boldsymbol{B}$. More precisely, $\mathbb{F}_2 \cdot F(B_1) = (0, B_1)$ and by induction, $\mathbb{F}_2 \cdot F(B_1, \ldots, B_i) = \mathbb{F}_2 \cdot F(B_1, \ldots, B_{i-1}) \| (B_i \otimes \mathbb{1}_{2^\gamma} + \mathbb{F}_2 \cdot F(B_1, \ldots, B_{i-1}))$. For instance, $\mathbb{F}_2 \cdot F(B_1, B_2, B_3) = (0, B_1, B_2, B_1 + B_2, B_3, B_1 + B_3, B_2 + B_3, B_1 + B_2 + B_3)$.

- Thanks to the shortening of $\mathscr{D}$ and the puncturing of $\mathscr{N}$ we are able to even more reduce the number of unknowns because for any $I \subset [\![1, n]\!]$ it holds

$$\mathcal{S}_I(\mathscr{D}) \star \mathcal{P}_I(\mathscr{N}) \subseteq \mathcal{S}_I(\mathscr{A}_r). \tag{9}$$

- Lastly, if the first $(k_0 - c)$ columns of $\boldsymbol{K}$ form an invertible matrix $\boldsymbol{S}$, we can then multiply the polynomial system by $\boldsymbol{S}^{-1}$ without altering the solution set. Therefore we may assume that the first columns of $\boldsymbol{K}$ forms the identity matrix, i.e. $\boldsymbol{K} = (\boldsymbol{I}_d \, \boldsymbol{U})$. Of course this observation also applies when considering the polynomial system defined in (9).

The algebraic attack harnesses (9) by first picking a set $I \subset [\![1, n]\!]$ of cardinality $2^\gamma a_0$ such that I is the union of a_0 disjoint dyadic blocks. The different steps are described below:

1. Recover $\mathcal{S}_I(\mathscr{D})$ and $\mathcal{P}_I(\mathscr{N})$ by solving the quadratic system

$$(\boldsymbol{I}_d \, \boldsymbol{U}) \, \mathcal{S}_I(\boldsymbol{G}_{\text{inv}}) \star \mathcal{P}_I(\boldsymbol{H}_{\text{pub}}) \cdot \mathcal{P}_I(\boldsymbol{V})^T = \boldsymbol{0} \tag{10}$$

where $d \triangleq \dim \mathcal{S}_I(\mathscr{D}) = k_0 - c - a_0$ and $\boldsymbol{V} = \boldsymbol{T} \otimes \mathbb{1}_{2^\gamma} + \mathbb{1}_{n_0} \otimes (\mathbb{F}_2 \cdot \boldsymbol{B})$.

2. Reconstruct $\mathcal{P}_I(\boldsymbol{x})$ from $\mathcal{S}_I(\mathscr{D})$ and $\mathcal{P}_I(\mathscr{N})$.

3. Recover $\mathcal{P}_I(\boldsymbol{y})$ then $\boldsymbol{x}$ and $\boldsymbol{y}$ using linear algebra.

Remark 4. Because of the particular form of the generator matrix of $\mathcal{S}_I(\mathscr{D})$ in (10), the system may have only a trivial solution. In other words, the polynomial system will provide $\mathcal{S}_I(\mathscr{D})$ if it admits a generator matrix such that its first d columns form an invertible matrix, which holds with probability $\prod_{i=1}^d (1 - q^{-i})$.

Remark 5. The attack searches for a code $\mathscr{D} \subset \mathscr{A}_r^{\mathrm{G}}$ of dimension $k_0 - c$. When $c \geqslant k_0$, like DAGS-3.1 where $k_0 = c = 16$, this code does not exist. But it is possible to use $\mathscr{A}_r^\perp$ instead of $\mathscr{A}_r$ to search for a code $\mathscr{E} \subset (\mathscr{A}_r^\perp)^{\mathrm{G}}$ of dimension $n_0 - k_0 - c$ such that

$$\mathscr{E} \star \mathscr{N} \subset \mathscr{A}_r^\perp \tag{11}$$

We will not present this version here because it does not give practical improvements.

The authors in [8] addressed Steps 2 and 3 by showing that it relies only on linear algebra with matrices of size at most n and can be done in $O(n^3)$ operations. In this paper, we are mainly interested in Step 1 and the computation of a Gröbner basis of (10).

Solving (10) Using Gröbner Bases. Using Propositon 1 and the fact that an affine map preserves the quasi-dyadic structure, we can assume that $b_1 = 1$ and $\tau_{n_0} = 0$. Moreover, any vector in the code $\mathcal{P}_I(\mathcal{N})$ is a solution to the system, in particular $\mathrm{T}_{\mathbb{F}_{q^2}/\mathbb{F}_q}(b_2)^{-1}\mathrm{T}_{\mathbb{F}_{q^2}/\mathbb{F}_q}(\mathcal{P}_I(x))$, so we know that a solution with $B_2 = 1$ exists.

Remark 6. Note that $\mathrm{T}_{\mathbb{F}_{q^2}/\mathbb{F}_q}(b_2)$ is not invertible if $b_2 \in \mathbb{F}_q$, which arises with probability $\frac{1}{q}$, but in this case a solution with $B_2 = 0, B_3 = 1$ exists in the system and we may specialize one more variable.

The following theorem identifies in (10) the number of equations and variables.

Theorem 2. *If G_{inv} and H_{pub} are in systematic form and $I = [\![1, a_0 2^\gamma]\!]$ then the polynomial system (10) with $T_{n_0} = 0, B_1 = 0, B_2 = 1$ contains*

$$
\begin{cases}
n_U = (k_0 - c - a_0)\,c & \textit{variables in } U \\
n_T = n_0 - k_0 + c - 1 & \textit{variables } T_{k_0-c+1}, \ldots, T_{n_0-1} \\
n_B = \gamma - 2 & \textit{variables } B_3, \ldots, B_\gamma \\
(k_0 - c - a_0)\,(n_0 - k_0 - 1) & \textit{quadratic equations}
\end{cases}
\tag{12}
$$

that are bilinear in the variables U and the variables in V, as well as $k_0 - c - a_0$ equations of the form

$$
T_i = P_i(U[i], T_{n_0-k_0+1}, \ldots, T_{n_0--1}, B), \quad i \in [\![1, k_0 - c - a_0]\!]
$$

where P_i is a bilinear polynomial in the variables U and V.

Proof. The hypothesis that $V = T \otimes \mathbb{1}_{2^\gamma} + \mathbb{1}_{n_0} \otimes (\mathbb{F}_2 \cdot B)$ reduces the number of variables, the number of solutions of the system (it restricts to quasi-dyadic solutions), but it also reduces the number of equations in the system by a factor 2^γ. Indeed, if we consider two rows c and c^σ from H_{pub} in the same quasi-dyadic block, where $\sigma \in \mathbb{G}$, then for any row $u = u^\sigma$ from the invariant matrix $(I_d\ U)\,\mathcal{S}_I(G_{\mathsf{inv}})$, the component-wise product with c^σ satisfies $u \star c^\sigma = (u \star c)^\sigma$ and the resulting equations are $(u \star c)^\sigma \cdot V^T = (u \star c \cdot V^T)^\sigma = u \star c \cdot V^T$ as $V = V^\sigma + \sigma \mathbb{1}_n$ and $u \star c \cdot \mathbb{1}_n = 0$.

The i-th row of $(I_d\ U)\,G_{\mathsf{inv}}$ contributes to $n_0 - k_0$ equations that contain the variables in the i-row $U[i]$ of U, B and $T_i, T_{k_0-c+1}, \ldots, T_{n_0-1}$. Moreover, by using the fact that the matrices are in systematic form, the component-wise product of this row by the i-th row of H_{pub} (the row in the i-th block as we take only one row every 2^γ rows) gives a particular equation that expresses T_i in terms of $T_{n_0-k_0+1}, \ldots, T_{n_0-1}$, B and $U[i]$. $\square$

5   Experimental Results

This section is devoted to the experimental results we obtained for computing a Gröbner basis. We consider two approaches for solving (10). The first one consists in solving the system without resorting to the shortening of codes ($I = \emptyset$). The second one treats the cases where we solve the system by shortening on I with different cardinalities.

We report the tests we have done using Magma [13] on a machine with a Intel® Xeon® 2.60 GHz processor. We indicate in the tables the number of clock cycles of the CPU, given by Magma using the `ClockCycles()` function, as well as the time taken on our machine.

Solving (10) without Shortening. The polynomial systems for the original DAGS parameters (DAGS-1, DAGS-3 and DAGS-5) are so overdetermined that it is possible to compute directly a Gröbner basis of (10) without shortening ($a_0 = 0$ *i.e* $I = \emptyset$). Table 3 gives the number of variables and equations for the DAGS parameters. One can see that there are at least 3 times as many equations as variables for the original parameters.

Moreover, the complexity of computing Gröbner bases is related to the highest degree of the polynomials reached during the computation. A precise analysis has been done for generic overdetermined homogeneous systems (called "semi-regular" systems) in [4,5] and for the particular case of generic bilinear systems in [22]. For generic overdetermined systems, this degree decreases when the number of polynomials increases. For DAGS-1, DAGS-3 and DAGS-5, the highest degree is small (3 or 4). Linear equations then appear at that degree which explains why we are able to solve the systems, even if the number of variables is quite large (up to 119 variables for DAGS-1).

Table 3. Computation time for the Gröbner basis of (10) without shortening $\mathscr{D}$ ($I = \emptyset$). The columns "dim $\mathscr{D}$" and "c" correspond to the dimension of $\mathscr{D}$ and the value of $c = \frac{q}{2^{\gamma}-1}$. The columns n_U and $n_V = n_B + n_T$ give the number of variables U and $B \cup T$ respectively. The column "Var." is equal to the total number of variables $n_U + n_V$ while the column "Eq." indicates the number of equations in (10). "Ratio" is equal to the ratio of the number of equations to the number of variables. "Gröb." gives the number of CPU clock cycles as given by the `ClockCycles()` function in Magma on Intel® Xeon® 2.60 GHz processor. "Deg." gives the degree where linear equations are produced. The column "Mat. size" is the size of the biggest matrix obtained during the computations.

Param.	dim $\mathscr{D}$	c	n_U	n_V	Var.	Eq.	Ratio	Gröb.	Deg.	Mat. size
DAGS-1	22	4	88	31	119	550	4.6	2^{44}	3	314,384 × 401,540
DAGS-3	12	4	48	28	76	252	3.3	2^{44}	4	725,895 × 671,071
DAGS-5	9	2	18	27	45	189	4.2	2^{33}	3	100,154 × 8,019

Solving (10) with Shortening. Shortening(10) on the i-th dyadic block consists exactly in selecting a subset of the system that does not contain variables from $U[i]$. If we are able to shorten the system without increasing the degree where linear equations appear, then the Gröbner basis computation is faster because the matrices are smaller.

For each set of parameters and for different dimensions of $\mathcal{S}_I(\mathscr{D})$, we ran 100 tests. The results are shown in Table 4. For these tests we always assumed that $b_2 \notin \mathbb{F}_q$ (see Remark 6 for more details). We recall that when $b_2 \in \mathbb{F}_q$ we may specialize one more variable, that is to say we take $B_1 = 0$, $B_2 = 0$ and $B_3 = 1$, and we are able to solve the system. In Table 4 we can see that the best results are obtained with $\dim \mathcal{S}_I(\mathscr{D}) = 4$ for DAGS-1, even if the number of variables is not the lowest. This can be explained for DAGS-1 by the fact that highest degree is 3 while when $\dim \mathcal{S}_I(\mathscr{D}) = 2$ or 3, the highest degree is 4.

The figures obtained for DAGS-3 show that the highest degree is 4 for any value of $\dim \mathcal{S}_I(\mathscr{D}) \geqslant 3$. But when $\dim \mathcal{S}_I(\mathscr{D}) = 4$ more linear equations are produced because the ratio of the number of equations to the number of variables is larger. With a $\dim \mathcal{S}_I(\mathscr{D}) = 2$, as Barelli and Couvreur [8] used in their attack, the ratio is small (1.17), and the maximal degree reached during the Gröbner basis computation is too large ($\geqslant 6$) to get a reasonable complexity. We had to stop because of a lack of memory and there was no linear equation at this degree.

For DAGS-5 the value of $\dim \mathcal{S}_I(\mathscr{D})$ has less influence on the performances because the system always has linear equations in degree 3.

Table 4. Computation time for the Gröbner basis in (10) with the shortening of $\mathscr{D}$. The column "Gröb." gives the number of CPU clock cycles as given by the ClockCycles() function in Magma on Intel® Xeon® 2.60 GHz processor.

Param.	$\dim \mathcal{S}_I(\mathscr{D})$	Var.	Eq.	Ratio	Gröb.	Time	Mem. (GB)	Deg.	Mat. size
DAGS-1	2	39	50	1.28	2^{39}	276 s	2.21	4	76,392 × 62,518
	3	43	75	1.74	2^{38}	163 s	1.11	4	97,908 × 87,238
	4	47	100	2.13	2^{33}	4 s	0.12	3	11,487 × 9,471
	5	51	125	2.45	2^{34}	6 s	0.24	3	11,389 × 13,805
DAGS-3	2	36	42	1.17	–	–	$\geqslant$139	$\geqslant$6	–
	3	40	63	1.58	2^{39}	321 s	1.24	4	85,981 × 101,482
	4	44	84	1.91	2^{37}	70 s	1.11	4	103,973 × 97,980
	5	48	105	2.19	2^{38}	140 s	1.48	4	170,256 × 161,067
DAGS-5	2	31	42	1.35	2^{31}	0.4 s	0.12	3	6,663 × 4,313
	3	33	63	1.91	2^{31}	0.4 s	0.13	3	4,137 × 4,066
	4	35	84	2.40	2^{31}	0.5 s	0.15	3	5,843 × 4,799
	5	37	105	2.84	2^{31}	0.4 s	0.19	3	6,009 × 4,839

Updated DAGS Parameters. After the publication of the attack in [8], the authors of DAGS proposed new parameters on their website [2]. They are given in Table 2. The computation of the Gröbner basis is no longer possible with this new set of parameters. For DAGS-1.1, the number of variables is so high that the computation involves a matrix in degree 4 with about two millions of rows, and we could not perform the computation. For DAGS-3.1, as $c = k_0$, the code $\mathscr{D}$ does not exist. Even by considering the dual of the public code as suggested by Remark 5, it was not feasible to solve the system. This is due to the fact the system is underdetermined: the ratio is 0.7 as shown in Table 5. As for DAGS-5.1, the system has too many variables and the ratio is too low.

Table 5. Updated parameters. Columns are the same as Table 3

Param.	dim $\mathscr{D}$	c	n_U	n_V	Var.	Eq.	Ratio
DAGS-1.1	18	8	144	35	179	450	2.5
DAGS-3.1	0	16	–	–	–	0	0
DAGS-3.1 dual	6	16	96	34	130	90	0.7
DAGS-5.1	12	16	192	40	232	252	1.1

However, as the systems are bilinear, a simple approach consisting in specializing a set of variables permits to get linear equations. For instance, with the new parameters for DAGS-1.1, and according to Theorem 2, the algebraic system contains $k_0 - c = \dim \mathscr{D} = 18$ sets of $n_0 - k_0 - 1 = 25$ equations bilinear in $c = 8$ variables U and $n_0 - k_0 + c + \gamma - 3 = 35$ variables V. If we specialize the U variables in a set of equations, we get 25 linear equations in 35 variables V. As $q = 2^6$ for DAGS-1.1, specializing 16 variables U gives a set of 50 linear equations in 35 variables that can be solved in at most $35^3 < 2^{15.39}$ finite field operations, and breaking DAGS-1.1 requires to test all values of the 16 variables U in $\mathbb{F}_q$, hence 2^{96} specializations, leading to an attack with $2^{111.39}$ operations, below the 128-bit security claim. This new set of parameters for DAGS-1.1 clearly does not take into account this point.

The next section will show how to cryptanalyze efficiently the DAGS-1.1 parameters.

6 Hybrid Approach on DAGS V2

We will show in this section that a hybrid approach mixing exhaustive search and Gröbner basis provides an estimated work factor of 2^{83} for DAGS-1.1.

As the algebraic system is still highly overdetermined with a ratio of 2.5 between the number of variables and the number of equations, we can afford to reduce the number of variables by shortening $\mathscr{D}$ over a_0 dyadic blocks while keeping a ratio large enough. For instance if $a_0 = k_0 - c - 2$ we have $\dim \mathcal{S}_I(\mathscr{D}) = 2$ and a system of 50 bilinear equations in 51 variables. On the other hand, specializing some variables as in the hybrid approach from [12] permits to increase the ratio. For each value of the variables we compute a Gröbner basis of the specialized system. When the Gröbner basis is $\langle 1 \rangle$, it means that the system has no solution, and it permits to "cut branches" of the exhaustive search. Experimentally, computations are quite fast for the wrong guesses, because the Gröbner basis computation stops immediately when 1 is found in the ideal.

Moreover, if we specialize an entire row of c variables U, as the equations are bilinear in U and V, then we get $n_0 - k_0 - a_0$ linear equations in V, which reduces the number of variables. Table 6 gives the complexity of the Gröbner basis computation for DAGS1.1, for $c = 8$ variables U specialized and different values of $\dim S_I(\mathscr{D})$.

Table 6. Experimental complexity of Gröbner basis computations with $c = 8$ specializations on U for DAGS1.1. The system after specialization contains "Var" remaining variables. The column "Linear" (resp. "Bilinear") gives the number of linear (resp. bilinear) equations in the system. "False" contains the complexity of the Gröbner basis computation for a wrong specialization, "True" for a correct one. The last column gives the global complexity of the attack if we have to test all possible values of the variables in $\mathbb{F}_q = \mathbb{F}_{2^6}$, that is $2^{6 \times 8} \times$ False + True.

$\dim S_I(\mathscr{D})$	Var	Linear	Bilinear	False	True	Total
2	43	25	25	2^{35}	2^{36}	2^{83}
3	51	25	50	2^{35}	2^{36}	2^{83}
4	59	25	75	2^{38}	2^{39}	2^{86}
5	67	25	100	2^{40}	2^{40}	2^{88}

Acknowledgements. This work has been supported by the French ANR projects MANTA (ANR-15-CE39-0013) and CBCRYPT (ANR-17-CE39-0007). The authors are extremely grateful to Élise Barelli for kindly giving her Magma code and for helpful discussions.

References

1. Banegas, G., et al.: DAGS: key encapsulation using dyadic GS codes. J. Math. Cryptology **12**(4), 221–239 (2018)
2. Banegas, G., et al.: DAGS: Key encapsulation for dyadic GS codes, specifications v2, September 2018
3. Banegas, G., et al.: DAGS: Key encapsulation for dyadic GS codes, November 2017. https://csrc.nist.gov/CSRC/media/Projects/Post-Quantum-Cryptography/documents/round-1/submissions/DAGS.zip. First round submission to the NIST post-quantum cryptography call
4. Bardet, M., Faugère, J.C., Salvy, B.: On the complexity of Gröbner basis computation of semi-regular overdetermined algebraic equations. In: International Conference on Polynomial System Solving, ICPSS 2004, 24–26 November, Paris, France, pp. 71–75 (2004)

5. Bardet, M., Faugère, J.C., Salvy, B., Yang, B.Y.: Asymptotic behaviour of the degree of regularity of semi-regular quadratic polynomial systems. In: MEGA 2005 Eighth International Symposium on Effective Methods in Algebraic Geometry, Porto Conte, Alghero, Sardinia, Italy, p. 15, 27 May–1 June 2005

6. Bardet, M., et al.: BIGQUAKE, November 2017. https://bigquake.inria.fr. NIST Round 1 submission for Post-Quantum Cryptography

7. Barelli, È.: On the security of some compact keys for McEliece scheme. In: WCC Workshop on Coding and Cryptography, September 2017

8. Barelli, É., Couvreur, A.: An efficient structural attack on NIST submission DAGS. In: Peyrin, T., Galbraith, S. (eds.) ASIACRYPT 2018. LNCS, vol. 11272, pp. 93–118. Springer, Cham (2018). https://doi.org/10.1007/978-3-030-03326-2_4

9. Berger, T.P., Cayrel, P.-L., Gaborit, P., Otmani, A.: Reducing key length of the McEliece cryptosystem. In: Preneel, B. (ed.) AFRICACRYPT 2009. LNCS, vol. 5580, pp. 77–97. Springer, Heidelberg (2009). https://doi.org/10.1007/978-3-642-02384-2_6

10. Bernstein, D.J., et al.: Classic McEliece: conservative code-based cryptography, November 2017. First round submission to the NIST post-quantum cryptography call

11. Bernstein, D.J., Lange, T., Peters, C.: Wild McEliece. In: Biryukov, A., Gong, G., Stinson, D.R. (eds.) SAC 2010. LNCS, vol. 6544, pp. 143–158. Springer, Heidelberg (2011). https://doi.org/10.1007/978-3-642-19574-7_10

12. Bettale, L., Faugère, J.C., Perret, L.: Hybrid approach for solving multivariate systems over finite fields. J. Math. Cryptology **3**(3), 177–197 (2009)

13. Bosma, W., Cannon, J., Playoust, C.: The magma algebra system I: the user language. J. Symbolic Comput. **24**(3/4), 235–265 (1997)

14. Couvreur, A., Gaborit, P., Gauthier-Umaña, V., Otmani, A., Tillich, J.P.: Distinguisher-based attacks on public-key cryptosystems using Reed-Solomon codes. Des. Codes Crypt. **73**(2), 641–666 (2014)

15. Couvreur, A., Otmani, A., Tillich, J.P.: Polynomial time attack on wild McEliece over quadratic extensions. In: Nguyen, P.Q., Oswald, E. (eds.) EUROCRYPT 2014. LNCS, vol. 8441, pp. 17–39. Springer, Heidelberg (2014). https://doi.org/10.1007/978-3-642-55220-5_2

16. Couvreur, A., Otmani, A., Tillich, J.P.: Polynomial time attack on wild McEliece over quadratic extensions. IEEE Trans. Inform. Theory **63**(1), 404–427 (2017)

17. Faugère, J.C., Otmani, A., Perret, L., de Portzamparc, F., Tillich, J.P.: Structural weakness of compact variants of the McEliece cryptosystem. In: 2014 Proceedings IEEE International Symposium Information Theory - ISIT, Honolulu, pp. 1717–1721, July 2014

18. Faugère, J.C., Otmani, A., Perret, L., de Portzamparc, F., Tillich, J.P.: Folding alternant and Goppa Codes with non-trivial automorphism groups. IEEE Trans. Inform. Theory **62**(1), 184–198 (2016)

19. Faugère, J.C., Otmani, A., Perret, L., de Portzamparc, F., Tillich, J.P.: Structural cryptanalysis of McEliece schemes with compact keys. Des. Codes Crypt. **79**(1), 87–112 (2016)

20. Faugère, J.-C., Otmani, A., Perret, L., Tillich, J.-P.: Algebraic cryptanalysis of McEliece variants with compact keys. In: Gilbert, H. (ed.) EUROCRYPT 2010. LNCS, vol. 6110, pp. 279–298. Springer, Heidelberg (2010). https://doi.org/10.1007/978-3-642-13190-5_14

21. Faugère, J.-C., Perret, L., de Portzamparc, F.: Algebraic attack against variants of McEliece with Goppa polynomial of a special form. In: Sarkar, P., Iwata, T. (eds.) ASIACRYPT 2014. LNCS, vol. 8873, pp. 21–41. Springer, Heidelberg (2014). https://doi.org/10.1007/978-3-662-45611-8_2

22. Faugère, J.C., El Din, M.S., Spaenlehauer, P.J.: Gröbner bases of bihomogeneous ideals generated by polynomials of bidegree (1,1): algorithms and complexity. J. Symbolic Comput. **46**(4), 406–437 (2011)

23. Gaborit, P.: Shorter keys for code based cryptography. In: Proceedings of the 2005 International Workshop on Coding and Cryptography (WCC 2005), pp. 81–91. Bergen, March 2005

24. Gauthier, V., Otmani, A., Tillich, J.P.: A distinguisher-based attack of a homomorphic encryption scheme relying on Reed-Solomon codes. CoRR abs/1203.6686 (2012)

25. Gauthier, V., Otmani, A., Tillich, J.P.: A distinguisher-based attack on a variant of McEliece's cryptosystem based on Reed-Solomon codes. CoRR abs/1204.6459 (2012)

26. McEliece, R.J.: A Public-Key System Based on Algebraic Coding Theory, pp. 114–116. Jet Propulsion Lab (1978). dSN Progress Report 44

27. Misoczki, R., Barreto, P.: Compact McEliece keys from Goppa codes. In: Selected Areas in Cryptography. Calgary, Canada, 13–14 August 2009

28. Otmani, A., Kalachi, H.T.: Square code attack on a modified Sidelnikov cryptosystem. In: El Hajji, S., Nitaj, A., Carlet, C., Souidi, E.M. (eds.) C2SI 2015. LNCS, vol. 9084, pp. 173–183. Springer, Cham (2015). https://doi.org/10.1007/978-3-319-18681-8_14

29. Wieschebrink, C.: Cryptanalysis of the Niederreiter public key scheme based on GRS subcodes. In: Sendrier, N. (ed.) PQCrypto 2010. LNCS, vol. 6061, pp. 61–72. Springer, Heidelberg (2010). https://doi.org/10.1007/978-3-642-12929-2_5

Weak Keys in the Faure–Loidreau Cryptosystem

Thomas Jerkovits[(✉)] and Hannes Bartz[(✉)]

Institute of Communication and Navigation, Deutsches Zentrum für Luft- und
Raumfahrt (DLR), 82234 Wessling, Germany
{thomas.jerkovits,hannes.bartz}@dlr.de

Abstract. Some types of weak keys in the Faure–Loidreau (FL) cryptosystem are presented. We show that from such weak keys the private key can be reconstructed with a computational effort that is substantially lower than the security level ($\approx 2^{25}$ operations for 80-bit security). The proposed key-recovery attack is based on ideas of generalized minimum distance (GMD) decoding for rank-metric codes.

Keywords: Code-based cryptography · Rank-metric codes ·
Interleaving · Gabidulin codes ·
Generalized minimum distance (GMD) decoding ·
Post-quantum cryptography · Faure–Loidreau

1 Introduction

Most current public-key cryptosystems like Rivest, Shamir and Adleman (RSA) [1] rely on hard mathematical problems such as prime factorization problem or the discrete logarithm problem. In 1999, Shor presented an algorithm for quantum computers that is able to solve the prime factorization problem and the discrete logarithm problem in polynomial time [2]. Thus, assuming that quantum computer of sufficient scale can be built one day, current cryptosystems like RSA can be broken in polynomial time rendering most of today's communication systems insecure. Current post-quantum secure public-key cryptosystems, i.e. systems that are resilient against attacks on quantum computers, suffer from large public keys compared to RSA, that means typically several hundreds of thousands of bits [3]. For example, the first code-based cryptosystem by McEliece [4] uses as a public key an obfuscated generator matrix of a linear block code, which results in a key size that is quadratic in the length of the code.

In 2006, Faure and Loidreau proposed a cryptosystem [5] that is based on the problem of reconstructing linearized polynomials. The Faure–Loidreau (FL) cryptosystem is the rank-metric equivalent of the Augot–Finiasz cryptosystem [6] and admits very small public keys for a given security level (≈ 2 KB for 80-bit security). In 2018, Gaborit et al. showed, that the private key in the FL cryptosystem can be recovered in polynomial time from the public key with high

M. Baldi et al. (Eds.): CBC 2019, LNCS 11666, pp. 102–114, 2019.
https://doi.org/10.1007/978-3-030-25922-8_6

probability. In [7] it was shown, that the attack from [8] is equivalent to the problem of list decoding interleaved Gabidulin codes [9]. In other words, the private key is a noisy codeword of an interleaved Gabidulin code with error weight chosen slightly larger then the unique decoding radius. Such noisy codewords can be recovered with a high probability by applying list decoding. That means the size of the list returned by the decoder is one with high probability. Whenever the list size is larger than one, the decoder fails. This kind of decoding is called probabilist unique decoding. By restricting to error patterns that make the probabilistic unique decoder of an interleaved Gabidulin decoder fail, the FL cryptosystem can be repaired [7].

In this paper, we consider a new method to recover the private key from the public key in the FL cryptosystem that uses properties that are *not* related to the previous key-recovery attacks in [7,8]. The method uses ideas from generalized minimum distance (GMD) decoding [10] which improves decoding by trading errors for erasures and also was applied for rank-metric codes [10,11]. This allows to recover private keys from some weak public keys with a computational complexity that is substantially below the security level of the cryptosystem. We characterize some types of weak keys and show that the key-recovery attack is feasible for the parameters suggested in [5,7].

2   Preliminaries

Let q be a power of a prime and denote by $\mathbb{F}_q$ a finite field of order q and by $\mathbb{F}_{q^m}$ the extension field of $\mathbb{F}_q$ of degree m. For an integer $u > 1$, we define an extension field $\mathbb{F}_{q^{mu}}$ of $\mathbb{F}_{q^m}$ such that $\mathbb{F}_q \subset \mathbb{F}_{q^m} \subset \mathbb{F}_{q^{mu}}$. By $\mathcal{B} = (\beta_1, \beta_2, \ldots, \beta_u)$ we denote an ordered basis of $\mathbb{F}_{q^{mu}}$ over $\mathbb{F}_{q^m}$.

We denote the set of all $m \times n$ matrices over $\mathbb{F}_q$ by $\mathbb{F}_q^{m \times n}$ and define $\mathbb{F}_q^n \overset{\text{def}}{=} \mathbb{F}_q^{1 \times n}$. Matrices and vectors are denoted by bold uppercase and lowercase letters such as $\mathbf{A}$ and $\mathbf{a}$, respectively. The Hamming weight $w_{\mathrm{H}}(\mathbf{a})$ of a vector $\mathbf{a}$ is defined as the number of nonzero entries in $\mathbf{a}$.

Under a fixed basis of $\mathbb{F}_{q^m}$ over $\mathbb{F}_q$ there is a bijective mapping between any vector $\mathbf{a} \in \mathbb{F}_{q^m}^n$ and a corresponding matrix $\mathbf{A} \in \mathbb{F}_q^{m \times n}$. The rank $\mathrm{rk}_q(\mathbf{a})$ of a vector $\mathbf{a} \in \mathbb{F}_{q^m}^n$ is defined as the rank of the corresponding matrix $\mathbf{A} \in \mathbb{F}_q^{m \times n}$ such that $\mathrm{rk}_q(\mathbf{a}) \overset{\text{def}}{=} \mathrm{rk}_q(\mathbf{A})$. The field trace of any $a \in \mathbb{F}_{q^{mu}}$ to $\mathbb{F}_{q^m}$ is denoted by $\mathrm{Tr}_{q^{mu}/q^m}(a)$. We use $\mathrm{Tr}_{q^{mu}/q^m}(\mathbf{a})$ to denote the element-wise field trace of a vector $\mathbf{a} \in \mathbb{F}_{q^{mu}}^n$.

For a given vector $\mathbf{a} \in \mathbb{F}_{q^m}^n$ and integer r, the *Moore* matrix is defined as

$$\mathbf{M}_r(\mathbf{a}) \overset{\text{def}}{=} \begin{pmatrix} a_1 & a_2 & \cdots & a_n \\ a_1^{[1]} & a_2^{[1]} & \cdots & a_n^{[1]} \\ \vdots & \vdots & \ddots & \vdots \\ a_1^{[r-1]} & a_2^{[r-1]} & \cdots & a_n^{[r-1]} \end{pmatrix}$$

where $[i] \overset{\text{def}}{=} q^i$ denotes the i-th Frobenius power.

2.1 Gabidulin Codes

The rank distance d_R between two matrices $\mathbf{A}, \mathbf{B} \in \mathbb{F}_q^{m \times n}$ with corresponding vectors $\mathbf{a}, \mathbf{b} \in \mathbb{F}_{q^m}^n$ is defined as

$$d_R(\mathbf{A}, \mathbf{B}) = d_R(\mathbf{a}, \mathbf{b}) \overset{\text{def}}{=} \text{rk}_q(\mathbf{A} - \mathbf{B}) = \text{rk}_q(\mathbf{a} - \mathbf{b}).$$

A linear rank-metric code of length $n \leq m$ and dimension k is an $\mathbb{F}_q$-linear subspace of $\mathbb{F}_q^{m \times n}$. The minimum rank distance d_R of a rank-metric code of length n and dimension k is upper bounded by the Singleton-like bound (see [12])

$$d_R \leq n - k + 1. \tag{1}$$

Codes that fulfill the Singleton-like bound in (1) with equality are called maximum rank distance (MRD) codes [12]. Gabidulin codes are a special class of rank-metric codes and fulfill (1) with equality, i.e. have minimum distance $d_R = n - k + 1$, and thus are MRD codes. A Gabidulin [12] code $\text{Gab}[n, k]$ of length n and dimension k over $\mathbb{F}_{q^m}$ is defined by the $\mathbb{F}_{q^m}$-linear row space of the generator matrix

$$\mathbf{G} = \mathbf{M}_k(\mathbf{g})$$

with $\mathbf{g} \in \mathbb{F}_{q^m}^n$ and $\text{rk}_q(\mathbf{g}) = n$.

A rank error channel takes a matrix $\mathbf{X} \in \mathbb{F}_q^{m \times n}$ as an input and outputs a matrix $\mathbf{Y} \in \mathbb{F}_q^{m \times n}$ such that

$$\mathbf{Y} = \mathbf{X} + \mathbf{E}$$

where the error matrix $\mathbf{E} \in \mathbb{F}_q^{m \times n}$ has rank $\text{rk}_q(\mathbf{E}) = t$.

Besides t rank errors, the transmitted matrix may be corrupted by row- or column erasures [13]. A matrix $\mathbf{X} \in \mathbb{F}_q^{m \times n}$ is corrupted by ρ row erasures and ϵ column erasures if ρ rows and ϵ columns are erased (i.e. set to zero). There exist efficient error erasure correcting decoders for Gabidulin codes that can correct t rank errors, ϵ column erasures and ρ row erasures up to

$$2t + \rho + \epsilon \leq n - k \tag{2}$$

requiring $\mathcal{O}(n^2)$ operations in $\mathbb{F}_{q^m}$ (see [14–16]). Note that ϵ column erasures in the matrix $\mathbf{X} \in \mathbb{F}_q$ correspond to ϵ symbol erasures in the corresponding vector $\mathbf{x} \in \mathbb{F}_{q^m}^n$.

2.2 Interleaved Gabidulin Codes

A homogeneous interleaved Gabidulin code $\text{IGab}[u; n, k]$ of length n, interleaving order u and component code dimension k over $\mathbb{F}_{q^m}$ is defined as the u-fold Cartesian product of a (component) Gabidulin code $\text{Gab}[n, k]$, i.e.

$$\text{IGab}[u; n, k] \overset{\text{def}}{=} \left\{ \begin{pmatrix} \mathbf{c}^{(1)} \\ \mathbf{c}^{(2)} \\ \vdots \\ \mathbf{c}^{(u)} \end{pmatrix} : \mathbf{c}^{(j)} \in \text{Gab}[n, k], \forall j = 1, \ldots, u \right\}.$$

An interleaved Gabidulin code $\text{IGab}[u; n, k]$ can correct rank errors up to

$$t \leq \frac{u}{u+1}(n-k)$$

with high probability (see e.g. [9,17,18]). Note, that interleaving improves the decoding radius for rank errors but does not improve the decoding radius for row and column erasures.

Under a fixed basis of $\mathbb{F}_{q^{mu}}$ over $\mathbb{F}_{q^m}$ a u-interleaved Gabidulin code $\text{IGab}[u; n, k]$ over $\mathbb{F}_{q^m}$ can be represented as the $\mathbb{F}_{q^{mu}}$-linear row space of the generator matrix $\mathbf{G} \in \mathbb{F}_{q^m}^{k \times n}$ of $\text{Gab}[n, k]$.

3   The Faure–Loidreau Cryptosystem

In the following a brief description of the FL cryptosystem is given. The encryption and decryption process are described in the Appendix. Let w be an integer that satisfies

$$\left\lfloor \frac{n-k}{2} \right\rfloor < w < n - k.$$

3.1   Key-Generation

1. Choose $\mathbf{g} \in \mathbb{F}_{q^m}^n$ with $\text{rk}_q(\mathbf{g}) = n$ at random and denote by $\mathbf{G} = \mathbf{M}_k(\mathbf{g})$ the generator matrix of the corresponding Gabidulin code $\text{Gab}[n, k]$.
2. Choose a vector $\mathbf{x} \in \mathbb{F}_{q^{mu}}^k$ such that the last u positions of $\mathbf{x}$ are $\mathbb{F}_{q^m}$-linearly independent at random.
3. Choose $\mathbf{s} = (s_1 \; s_2 \; \ldots \; s_w) \in \mathbb{F}_{q^{mu}}^w$ such that $\text{rk}_q(\mathbf{s}) = w$ at random.
4. Choose a random invertible matrix $\mathbf{P} \in \mathbb{F}_q^{n \times n}$ and compute

$$\mathbf{z} = (\, \mathbf{s} \,|\, \mathbf{0} \,) \, \mathbf{P}^{-1}. \tag{3}$$

Private key: $(\mathbf{P}, \mathbf{z}, \mathbf{x})$

Public key: $(\mathbf{g}, k, \mathbf{k}_{\text{pub}}, t_{\text{pub}})$ where

$$\mathbf{k}_{\text{pub}} = \mathbf{x}\mathbf{G} + \mathbf{z} \tag{4}$$

and

$$t_{\text{pub}} = \left\lfloor \frac{n-k-w}{2} \right\rfloor. \tag{5}$$

3.2   Key-Recovery Attacks on the Faure–Laudreau Cryptosystem

Gaborit, Otmani and Kalachi showed that an alternative private key $\mathbf{k}'_{\text{pub}}$ can be recovered in polynomial time from the public key $\mathbf{k}_{\text{pub}}$ and $\mathbf{G}$ with high probability if

$$w \leq \frac{u}{u+1}(n-k) \tag{6}$$

holds [8]. The attack cannot be prevented by adjusting the parameters, e.g. by choosing $w > \frac{u}{u+1}(n-k)$, since w reduces t_{pub} (see (5)) such that decoding attacks become feasible.

Consider the public key $\mathbf{k}_{\text{pub}}$ from (4) where $\mathbf{x} \in \mathbb{F}_{q^{mu}}^{k}$ and $\mathbf{z} \in \mathbb{F}_{q^{mu}}^{n}$ with $\text{rk}_q(\mathbf{z}) = w$. By defining $\mathbf{k}_{\text{pub}} = \sum_{i=1}^{u} \mathbf{k}_{\text{pub}}^{(i)}\beta_i$, $\mathbf{z} = \sum_{i=1}^{u} \mathbf{z}^{(i)}\beta_i$ and $\mathbf{x} = \sum_{i=1}^{u} \mathbf{x}^{(i)}\beta_i$ we can express the public key as

$$
\begin{pmatrix} \mathbf{k}_{\text{pub}}^{(1)} \\ \mathbf{k}_{\text{pub}}^{(2)} \\ \vdots \\ \mathbf{k}_{\text{pub}}^{(u)} \end{pmatrix} = \begin{pmatrix} \mathbf{x}^{(1)}\mathbf{G} \\ \mathbf{x}^{(2)}\mathbf{G} \\ \vdots \\ \mathbf{x}^{(u)}\mathbf{G} \end{pmatrix} + \begin{pmatrix} \mathbf{z}^{(1)} \\ \mathbf{z}^{(2)} \\ \vdots \\ \mathbf{z}^{(u)} \end{pmatrix} \tag{7}
$$

where $\mathbf{x}^{(j)} \in \mathbb{F}_{q^m}^{k}$ and $\mathbf{k}_{\text{pub}}^{(j)}, \mathbf{z}^{(j)} \in \mathbb{F}_{q^m}^{n}$ with $\text{rk}_q(\mathbf{z}^{(j)}) \leq w$ for all $j = 1, \ldots, u$. Note, that the public key in (7) is a codeword of an interleaved Gabidulin code $\text{IGab}[u; n, k]$ over $\mathbb{F}_{q^m}$ that is corrupted by an error of rank w.

Recently, Wachter-Zeh et al. showed that the key-recovery attack from [8] is equivalent to the decoding problem of an interleaved Gabidulin code [7, Thm. 3], i.e. to the problem of recovering $\mathbf{x}^{(1)}, \ldots, \mathbf{x}^{(u)}$ and $\mathbf{z}^{(1)}, \ldots, \mathbf{z}^{(u)}$ from (7). Hence, an attacker can reconstruct an alternative private key by running an interleaved decoder on the public key $\mathbf{k}_{\text{pub}}$, which is possible since the generator matrix $\mathbf{G}$ of the component codes is public.

This observation provides an explicit repair of the FL system by choosing the error vectors $\mathbf{z}^{(j)}$ for $j = 1, \ldots, u$ in the key generation step (see Sect. 3.1) such that a probabilistic unique interleaved Gabidulin decoder fails although w satisfies (6). The most secure choice for $\mathbf{z}$ in (7) is such that $\mathbf{z}^{(1)} = \mathbf{z}^{(2)} = \cdots = \mathbf{z}^{(u)}$ with $\text{rk}_q(\mathbf{z}^{(j)}) = w$ for all $j = 1, \ldots, u$. The number of keys is still large enough by restricting to error vectors of this form [7]. Hence, the cryptosystem can be repaired by constructing the component error vectors of $\mathbf{z}$ in (3) as

$$
\mathbf{z}^{(j)} = (\mathbf{s}^{(1)}|\mathbf{0})\mathbf{P}^{-1} \qquad \forall j = 1, \ldots, u
$$

with $\mathbf{s}^{(1)} \in \mathbb{F}_{q^m}^{w}$ and $\text{rk}_q(\mathbf{s}^{(1)}) = w$. The parameters of the repaired FL system as in [7] are given in Table 1.

4 A GMD-Based Key-Recovery Attack

In this section we present a new attack on the repaired FL cryptosystem that allows to recover the private key of some weak public keys efficiently. The attack is based on the principle of GMD decoding [10]. The general idea of GMD decoding is to incorporate soft information (e.g. from the communication channel) in the bounded minimum distance (BMD) decoding process by erasing the most unreliable positions. This results in an improved error correction performance, since a BMD decoder can correct twice more erasures than errors. The principle of GMD decoding can be extended to rank-metric codes [11], where rank errors can be traded for row and column erasures (see (2)). An erasure in the case of rank-metric codes can occur either row and/or column wise in the corresponding codeword matrix. That means the entries in one or more rows and/or columns are erased but the positions of the rows and columns are known.

Although an attacker is not provided with soft information about the error vector z that obfuscates the private information about the public key (4), we show that a GMD-based key-recovery attack that exploits the improved correction capability for row and column erasures is feasible for some error patterns z, even if all possible combinations of row and column erasure positions need to be considered. The knowledge of z allows to recover x by computing

$$x = (k_{pub} - z)G^{\dagger}$$

where $G^{\dagger}$ is the right inverse of the generator matrix G. Hence, an alternative private key $(\tilde{P}, x, z)$ can be obtained by computing an invertible matrix $\tilde{P} \in \mathbb{F}_q^{n \times n}$ satisfying

$$zP = \left(\tilde{s}^{(1)} | 0\right)$$

with $\tilde{s}^{(1)} \in \mathbb{F}_{q^m}^w$ and $\mathrm{rk}_q(\tilde{s}^{(1)}) = w$. Since one $z^{(j)}$ in (7) is sufficient to recover the whole error vector z in the repaired FL system, we focus on the first row of the expanded public key in (7), i.e.

$$k_{pub}^{(1)} = x^{(1)}G + z^{(1)} \quad \text{with } \mathrm{rk}_q(z^{(1)}) = w.$$

A straightforward decoding approach using a BMD Gabidulin decoder for the code $\mathrm{Gab}[n, k]$ with generator matrix G will fail since $w = \mathrm{rk}_q(z^{(1)})$ is chosen such that $w > (n - k)/2$. We define the excess of the error rank w over the unique decoding radius as

$$\xi \overset{\text{def}}{=} w - \frac{n - k}{2}. \tag{8}$$

Recall, that for Gabidulin codes there exist algorithms [13, 14, 16] that can correct $\delta = \epsilon + \rho$ row and column erasures and errors of rank w up to (see (2))

$$2w + \delta \leq n - k.$$

By artificially imposing δ row and column erasures on $k_{pub}^{(1)}$ and thus on the error vector $z^{(1)}$, one obtains a new error vector $z'^{(1)}$ with $w' \overset{\text{def}}{=} \mathrm{rk}_q(z'^{(1)}) \leq w$. Thus, an error and row/column erasure decoder can successfully decode if

$$2w' + \delta \leq n - k. \tag{9}$$

Using (8), (9) and the fact that $w - \delta \leq w' \leq w$, we obtain[1]

$$2\xi \leq \delta \leq n - k. \tag{10}$$

---

[1] Note, that by definition 2ξ is always an integer (see (8)).

Let $\mathsf{Dec}_{\mathbf{G}}(\cdot)$ denote an efficient error erasure decoding algorithm for the Gabidulin code $\text{Gab}[n, k]$ characterized by the generator matrix $\mathbf{G}$ that returns an estimate $\hat{\mathbf{x}}^{(1)}\mathbf{G}$ of the "transmitted" codeword $\mathbf{x}^{(1)}\mathbf{G} \in \text{Gab}[n, k]$.

We define the set $\mathcal{I}_\delta$ of row and column erasure patterns as

$$\mathcal{I}_\delta \overset{\text{def}}{=} \{(\mathbf{a}, \mathbf{b}) : \mathbf{a} \in \mathbb{F}_2^m, \mathbf{b} \in \mathbb{F}_2^n \text{ with } \mathsf{w}_H(\mathbf{a}) + \mathsf{w}_H(\mathbf{b}) = \delta\}$$

where the nonzero entries in $\mathbf{a}$ and $\mathbf{b}$ indicate the row and column erasure positions and $\mathsf{w}_H(\cdot)$ denotes the Hamming weight of a binary vector.

For any matrix $\mathbf{Y} \in \mathbb{F}_q^{m \times n}$ let $\mathcal{E}_{(\mathbf{a},\mathbf{b})}(\mathbf{Y})$ denote the row and column erasure operator that returns the matrix $\mathbf{Y} \in \mathbb{F}_q^{m \times n}$ where the rows and columns are erased (i.e. set to zero) according to the erasure pattern $(\mathbf{a}, \mathbf{b})$. We may also use the operator $\mathcal{E}_{(\mathbf{a},\mathbf{b})}(\mathbf{y})$ on the corresponding vector $\mathbf{y} \in \mathbb{F}_{q^m}^n$. Algorithm 1 summarizes the proposed GMD-based key-recovery attack.

Algorithm 1. A GMD-based Key-Recovery Attack

Input : $\mathbf{k}_{\text{pub}}^{(1)}, \mathbf{G}, w, \mathcal{I}_\delta, N_{\max}$

Output: $\hat{\mathbf{x}}^{(1)}, \hat{\mathbf{z}}^{(1)}$ with $\text{rk}_q(\hat{\mathbf{z}}^{(1)}) = w$ s.t. $\hat{\mathbf{x}}^{(1)}\mathbf{G} + \hat{\mathbf{z}}^{(1)} = \mathbf{k}_{\text{pub}}^{(1)}$ or "failure"

1 **foreach** $i \in [1, N_{max}]$ **do**
2   **foreach** $\delta \in [2\xi, n - k]$ **do**
3   Pick an erasure pattern $(\mathbf{a}, \mathbf{b})$ uniformly at random from $\mathcal{I}_\delta$
4        $\hat{\mathbf{c}}^{(1)} \leftarrow \mathsf{Dec}_{\mathbf{G}}\left(\mathcal{E}_{(\mathbf{a},\mathbf{b})}(\mathbf{k}_{\text{pub}})\right)$
5   **if** $\hat{\mathbf{c}}^{(1)} \neq \emptyset$ **then**
6           $\hat{\mathbf{x}}^{(1)} \leftarrow \hat{\mathbf{c}}^{(1)}\mathbf{G}^\dagger$
7           $\hat{\mathbf{z}}^{(1)} \leftarrow \mathbf{k}_{\text{pub}}^{(1)} - \hat{\mathbf{x}}^{(1)}\mathbf{G}$
8   **if** $\text{rk}_q(\hat{\mathbf{z}}^{(1)}) = w$ **then**
9   **return** $\hat{\mathbf{x}}^{(1)}, \hat{\mathbf{z}}^{(1)}$

10 **return** *"failure"*

In some cases the rank w of the error $\mathbf{z}^{(1)}$ is not reduced enough by the δ artificial row and column erasures on $\mathbf{k}_{\text{pub}}^{(1)}$ such that (9) is not satisfied. In this case we either get a miscorrection (i.e. decoder returns an estimate $\hat{\mathbf{c}}^{(1)} \neq \mathbf{c}^{(1)}$) or a decoding failure ($\mathsf{Dec}_{\mathbf{G}}(\cdot) = \emptyset$). Line 8 detects miscorrections that lead to codewords $\hat{\mathbf{c}}^{(1)}$ that are not at rank distance w from $\mathbf{k}_{\text{pub}}$.[2]

The worst-case computational complexity of the attack in Algorithm 1 is on the order of $|\mathcal{I}_\delta|\mathcal{O}(n^2)$ operations in $\mathbb{F}_{q^m}$. The codes considered for the FL cryptosystem are rather short (see Table 1). That means it is computationally

---

[2] There may be estimates $\hat{\mathbf{c}}^{(1)}$ at rank distance w from $\mathbf{k}_{\text{pub}}$ such that $\hat{\mathbf{c}}^{(1)} \neq \mathbf{c}^{(1)}$. However, this event is very unlikely since ξ is very small for the considered parameters (see Table 1) and was not observed in our simulations.

affordable to iterate through all possible erasure patterns with δ row and column erasures which are satisfying (10). Algorithm 1 can be parallalized to improve the runtime of the attack.

We say that the attack in Algorithm 1 is successful if the algorithm outputs $\hat{\mathbf{x}}^{(1)}, \hat{\mathbf{z}}^{(1)}$ (not "failure"). This means, that there exists an erasure pattern incorporating δ row and column erasures such that (9) is satisfied.

5   Classification of Weak Keys

In this section we classify some types of weak keys, in particular the corresponding error vectors $\mathbf{z}^{(1)}$, that are vulnerable against the key-recovery attack in Algorithm 1.

Let the invertible matrix $\mathbf{P}^{-1} \in \mathbb{F}_q^{n \times n}$ in (3) be partitioned such that

$$\mathbf{P}^{-1} = \begin{pmatrix} \mathbf{P}_1 \\ \mathbf{P}_2 \end{pmatrix}$$

with matrices $\mathbf{P}_1 \in \mathbb{F}_q^{w \times n}$ and $\mathbf{P}_2 \in \mathbb{F}_q^{(n-w) \times n}$ having full rank. Then by (3) we have that

$$\mathbf{z}^{(1)} = \mathbf{s}^{(1)} \mathbf{P}_1. \tag{11}$$

Since $\mathbf{P}_1$ is of full rank we can decompose it as

$$\mathbf{P}_1 = \mathbf{A} \cdot [\mathbf{I}_{w \times w} \,|\, \mathbf{Q}] \cdot \mathbf{B} \tag{12}$$

where $\mathbf{A} \in \mathbb{F}_q^{w \times w}$ is of full rank and $\mathbf{B} \in \mathbb{F}_q^{n \times n}$ is a permutation matrix.

In the following we restrict to column erasure attacks only (i.e. $\delta = \epsilon$) and describe weak keys by the structure of $\mathbf{Q}$. The decomposition in (12) of $\mathbf{P}_1$ is in general not unique. By "weak keys" we refer to error vectors $\mathbf{z}^{(1)}$ of the form (11) for which there exists a decomposition (12) of $\mathbf{P}_1$ with $\mathbf{Q}$ having a structure as described in the following subsections. Note, that by writing (11) over $\mathbb{F}_q$ we get similar arguments for row erasures on the rows of the corresponding matrix $\mathbf{S}^{(1)}$ of the vector $\mathbf{s}^{(1)}$.

5.1   Rank Equal to Hamming Weight Error Patterns

The rank of $\mathbf{z}^{(1)}$ is equal to its Hamming weight, i.e. $\mathrm{rk}_q(\mathbf{z}^{(1)}) = \mathrm{w}_H(\mathbf{z})$, if $\mathbf{Q}$ in (12) is the allzero matrix ($\mathbf{Q} = \mathbf{0}$). A necessary condition for the success of Algorithm 1 is that the erasure pattern for the column erasures at a certain iteration is chosen such that (9) holds. Let us denote by $S_{\epsilon,1}$ such an event for a specific number of column erasures ϵ given that the error vector $\mathbf{z}^{(1)}$ has rank equal to its Hamming weight. For this special case we can compute the success probability as

$$\Pr(S_{\epsilon,1}) = \frac{\displaystyle\sum_{i=\lceil \xi+\epsilon/2 \rceil}^{\min(\epsilon,w)} \binom{w}{i} \binom{n-w}{\epsilon-i}}{\binom{n}{\epsilon}} \tag{13}$$

where the denominator is the number of all possible column erasure patterns for a given ϵ and the numerator is the sum of the number of events for which i many errors are traded for erasures. The maximum number of errors that can be traded for erasures is $\min(\epsilon, w)$. The minimum amount of errors that need to be traded for erasures such that (9) holds is $\lceil \xi + \epsilon/2 \rceil$, which we obtain by inserting the relation $w' = w - i$ into (9) and using the definition of ξ from (8). The probability $\Pr(S_{\epsilon,1})$ for the 80-bit security parameters proposed in [7] is shown in Fig. 1.

5.2 η-Weak Error Patterns

The weakness of the keys described by an allzero matrix $\mathbf{Q}$ comes from the fact that whenever one of the w nonzero entries in $\mathbf{z}$ is hit by an erasure the rank of $\mathbf{z}$ is reduced by one. The number of these events is considerably high (see numerator of (13)), even for error patterns that can be decomposed as in (12) with a matrix $\mathbf{Q}$ having a certain amount of allzero rows. Let η denote the fraction of allzero rows in $\mathbf{Q}$, i.e. $\eta = N_0/w$, where N_0 is the number of allzero rows in $\mathbf{Q}$. Clearly, for $\eta = 1$ we have that $w_H(\mathbf{z}^{(1)}) = \mathrm{rk}_q(\mathbf{z}^{(1)})$. The success probability of Algorithm 1 for keys with $\eta < 1$, which we refer to as η-weak error patterns, depends on the remaining nonzero entries of $\mathbf{Q}$.

Based on (13) we derive a lower bound on the success probability by assuming that only the positions in $\mathbf{z}^{(1)}$ that are related to the allzero rows of $\mathbf{Q}$ are the cause for the rank reduction of $\mathbf{z}$. Let $S_{\epsilon,\eta}$ denote the event that in a certain iteration of Algorithm 1 an erasure pattern with ϵ columns erasures such that (9) holds is picked, given that $\mathbf{z}^{(1)}$ is an η-weak error pattern. The probability of $S_{\epsilon,\eta}$ can be lower bounded by

$$\Pr(S_{\epsilon,\eta}) \geq P(\epsilon, \eta) \overset{\text{def}}{=} \frac{\displaystyle\sum_{i=\lceil \xi+\epsilon/2 \rceil}^{\min(\epsilon, w\eta)} \binom{w\eta}{i} \binom{n - w\eta}{\epsilon - i}}{\binom{n}{\epsilon}} \tag{14}$$

that is the cumulative sum of a hypergeometric distribution.

The tightness of the lower bound $P(\epsilon, \eta)$ in (14) is validated by simulations for different η-weak error patterns $\mathbf{z}^{(1)}$ for the 80-bit security parameters from [7]. The simulation results are illustrated in Fig. 1. The success rates for the smallest possible η for different security levels are given in Table 1.

5.3 Further Weak Error Patterns

Simulation results show that there exist further error patterns that are different from the previously characterized patterns and can be recovered by Algorithm 1. In particular, another class of weak keys is characterized by matrices $\mathbf{Q}$ with $\mathrm{rk}_q(\mathbf{Q}) \ll \min(w, n - w)$. Figure 2 shows the simulations results of the success rate over ϵ for matrices $\mathbf{Q}$ with $\mathrm{rk}_q(\mathbf{Q}) = 1$ and $\mathrm{rk}_q(\mathbf{Q}) = 2$ that do not correspond to η error patterns with $\eta > 0$.

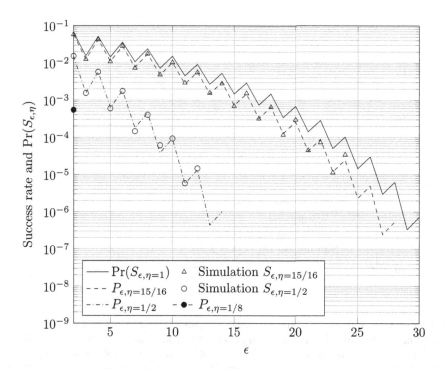

Fig. 1. Conditional success rate and success probability of a column erasure attack according to Algorithm 1 for η-weak error patterns with different values of η using the parameters proposed in [7] for the 80-bit security level. For each value of η and ϵ the simulation results were obtained by Monte Carlo Simulations running $2 \cdot 10^6$ iterations and choosing in every iteration a random η-weak error pattern $\mathbf{z}^{(1)}$ with $\mathrm{rk}_q(\mathbf{z}^{(1)}) = w$ and $\mathbf{Q}$ having $w\eta$ allzero rows.

Table 1. Parameters for the repaired FL cryptosystem in [7] and the corresponding probabilities for η-weak error patterns with $\eta = 2\xi/w$ and a column erasure attack with $\epsilon = 2\xi$ column erasures.

Security level	q	m	u	n	k	w	Key size	ξ	$\log_2(\mathrm{Pr}(S_{2\xi,2\xi/w}))$
80-bit	2	61	3	61	31	16	1.86 KB	1	−10.82
128-bit	2	63	3	63	31	18	1.98 KB	2	−19.16
256-bit	2	82	4	82	48	20	4.20 KB	3	−28.36

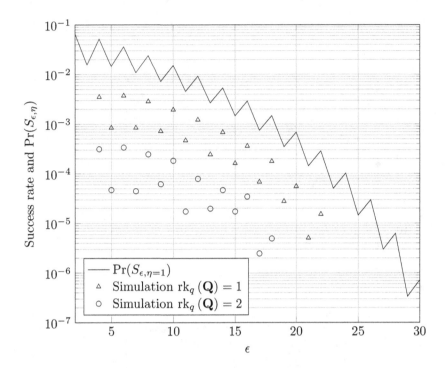

Fig. 2. Conditional success rate and success probability of a column erasure attack according to Algorithm 1 for weak keys with $\mathbf{Q}$ of low rank and having no allzero rows. The parameters proposed in [7] for the 80-bit security level are used. For each value of ϵ the simulation results were obtained by Monte Carlo simulations runing $2 \cdot 10^6$ iterations and choosing in every iteration a random error vector $\mathbf{z}^{(1)}$ with $\mathrm{rk}_q(\mathbf{z}^{(1)}) = w$ and $\mathrm{rk}_q(\mathbf{Q}) \in \{1, 2\}$. The probability of success $\Pr(S_{\epsilon, \eta=1})$ for error patterns with Hamming weight equal to their rank weight is plotted as a reference.

6 Conclusions

A new key-recovery attack on the Faure–Loidreau (FL) system was presented. The attack uses ideas from generalized minimum distance (GMD) decoding of rank-metric codes to recover private keys from some weak public keys. Some families of weak keys were classified and analyzed.

We showed, that for the current parameters, the private keys of some weak public keys can be recovered with an effort which is substantially lower than the security level of the cryptosystem. The classification of all weak keys that are vulnerable to the attack presented in this paper is still an open problem and subject to further research. That means we require that by generating a random public key as proposed in the FL cryptosystem the probability of picking a weak key that is affected by the attack is within the proposed security level.

A   Appendix

A.1   Encryption

Let $\mathbf{m} = (m_1\ m_2\ \ldots\ m_{k-u}\ |\ 0\ \ldots\ 0) \in \mathbb{F}_{q^m}^k$ be the plaintext.

1. Choose an element $\alpha \in \mathbb{F}_{q^{mu}}$ at random.
2. Choose $\mathbf{e} \in \mathbb{F}_{q^m}^n$ with $\mathrm{rk}_q(\mathbf{e}) \leq t_{\mathrm{pub}}$ at random.
3. Compute the ciphertext $\mathbf{c} \in \mathbb{F}_{q^m}^n$ as

$$\mathbf{c} = \mathbf{m}\mathbf{G} + \mathrm{Tr}_{q^{mu}/q^m}(\alpha\mathbf{k}_{\mathrm{pub}}) + \mathbf{e}.$$

A.2   Decryption

1. Compute

$$\mathbf{c}\mathbf{P} = \mathbf{m}\mathbf{G}\mathbf{P} + \mathrm{Tr}_{q^{mu}/q^m}(\alpha\mathbf{k}_{\mathrm{pub}})\mathbf{P} + \mathbf{e}\mathbf{P}.$$

Due to the $\mathbb{F}_{q^m}$-linearity of the trace we have

$$\mathrm{Tr}_{q^{mu}/q^m}(\alpha\mathbf{k}_{\mathrm{pub}})\mathbf{P} = \mathrm{Tr}_{q^{mu}/q^m}(\alpha\mathbf{x})\mathbf{G}\mathbf{P} + (\mathrm{Tr}_{q^{mu}/q^m}(\alpha\mathbf{s})\,|\,\mathbf{0})$$

and get

$$\mathbf{c}\mathbf{P} = (\mathbf{m} + \mathrm{Tr}_{q^{mu}/q^m}(\alpha\mathbf{x}))\mathbf{G}\mathbf{P} + (\mathrm{Tr}_{q^{mu}/q^m}(\alpha\mathbf{s})\,|\,\mathbf{0}) + \mathbf{e}\mathbf{P}.$$

2. Define $\mathbf{G}'_P$ as the last $n - w$ columns of the product $\mathbf{G}\mathbf{P}$ and let $\mathbf{c}'$ and $\mathbf{e}'$ be the last $n - w$ positions of $\mathbf{c}$ and $\mathbf{e}\mathbf{P}$, respectively. Then we have that

$$\mathbf{c}' = (\mathbf{m} + \mathrm{Tr}_{q^{mu}/q^m}(\alpha\mathbf{x}))\mathbf{G}'_P + \mathbf{e}'$$

with $\mathrm{rk}_q(\mathbf{e}') \leq t_{\mathrm{pub}} = \lfloor \frac{n-k-w}{2} \rfloor$.

Since $\mathbf{G}'_P$ is a generator matrix of $\mathrm{Gab}[n - w, k]$ we can decode to remove $\mathbf{e}'$ and get

$$\mathbf{m}' = \mathbf{m} + \mathrm{Tr}_{q^{mu}/q^m}(\alpha\mathbf{x}).$$

3. Since $\mathbf{m} = (m_1\ m_2\ \ldots\ m_{k-u}\ |\ \underbrace{0\ \ldots\ 0}_{u})$ we have that the last u positions of $\mathbf{m}'$ are

$$m'_i = \mathrm{Tr}_{q^{mu}/q^m}(\alpha x_i), \qquad \forall i = k - u + 1, \ldots, k.$$

Since $\mathcal{X} \stackrel{\mathrm{def}}{=} (x_{k-u+1}, \ldots, x_k)$ forms an ordered basis of $\mathbb{F}_{q^{mu}}$ over $\mathbb{F}_{q^m}$ we can compute α as

$$\alpha = \sum_{i=k-u+1}^{k} \mathrm{Tr}_{q^{mu}/q^m}(\alpha x_i)x_i^{\perp} = \sum_{i=k-u+1}^{k} m'_i x_i^{\perp},$$

where $\mathcal{X}^{\perp} \stackrel{\mathrm{def}}{=} (x_{k-u+1}^{\perp}, \ldots, x_k^{\perp})$ denotes the dual basis of $\mathcal{X}$. Finally, we can recover the plaintext as

$$\mathbf{m} = \mathbf{m}' - \mathrm{Tr}_{q^{mu}/q^m}(\alpha\mathbf{x}).$$

References

1. Rivest, R.L., Shamir, A., Adleman, L.: A method for obtaining digital signatures and public-key cryptosystems. Commun. ACM **21**(2), 120–126 (1978)
2. Shor, P.W.: Polynomial-time algorithms for prime factorization and discrete logarithms on a quantum computer. SIAM J. Comput. **26**(5), 1484–1509 (1997)
3. Loidreau, P.: Designing a rank metric based McEliece cryptosystem. In: Sendrier, N. (ed.) PQCrypto 2010. LNCS, vol. 6061, pp. 142–152. Springer, Heidelberg (2010). https://doi.org/10.1007/978-3-642-12929-2_11
4. McEliece, R.J.: A public-key cryptosystem based on algebraic codes. Deep Space Network Progress Report 44, pp. 114–116 (1978)
5. Faure, C., Loidreau, P.: A new public-key cryptosystem based on the problem of reconstructing p–polynomials. In: Ytrehus, Ø. (ed.) WCC 2005. LNCS, vol. 3969, pp. 304–315. Springer, Heidelberg (2006). https://doi.org/10.1007/11779360_24
6. Augot, D., Finiasz, M.: A public key encryption scheme based on the polynomial reconstruction problem. In: Biham, E. (ed.) EUROCRYPT 2003. LNCS, vol. 2656, pp. 229–240. Springer, Heidelberg (2003). https://doi.org/10.1007/3-540-39200-9_14
7. Wachter-Zeh, A., Puchinger, S., Renner, J.: Repairing the Faure-Loidreau public-key cryptosystem. In: IEEE International Symposium on Information Theory (ISIT) (2018)
8. Gaborit, P., Otmani, A., Kalachi, H.T.: Polynomial-time key recovery attack on the Faure-Loidreau scheme based on Gabidulin codes. Des. Codes Crypt. **86**(7), 1391–1403 (2018)
9. Loidreau, P., Overbeck, R.: Decoding rank errors beyond the error correcting capability. In: International Workshop on Algebraic and Combinatorial Coding Theory (ACCT), pp. 186–190, September 2006
10. Forney, G.: Generalized minimum distance decoding. IEEE Trans. Inf. Theory **12**(2), 125–131 (1966)
11. Bossert, M., Costa, E., Gabidulin, E., Schulz, E., Weckerle, M.: Verfahren und Kommunikationsvorrichtung zum Dekodieren von mit einem Rang-Code codierten Daten, EU Patent EP20 040 104 458 (2003)
12. Gabidulin, E.M.: Theory of codes with maximum rank distance. Problemy Peredachi Informatsii **21**(1), 3–16 (1985)
13. Gabidulin, E.M., Paramonov, A.V., Tretjakov, O.V.: Rank errors and rank erasures correction. In: 4th International Colloquium on Coding Theory (1991)
14. Richter, G., Plass, S.: Error and erasure decoding of rank-codes with a modified Berlekamp-Massey algorithm. In: International ITG Conference on Systems, Communications and Coding (SCC) (2004)
15. Gabidulin, E.M., Pilipchuk, N.I.: Error and erasure correcting algorithms for rank codes. Des. Codes Crypt. **49**(1–3), 105–122 (2008)
16. Silva, D.: Error control for network coding, Ph.D. dissertation (2009)
17. Sidorenko, V., Jiang, L., Bossert, M.: Skew-feedback shift-register synthesis and decoding interleaved Gabidulin codes. IEEE Trans. Inf. Theory **57**(2), 621–632 (2011)
18. Wachter-Zeh, A., Zeh, A.: List and unique error-erasure decoding of interleaved Gabidulin codes with interpolation techniques. Des. Codes Crypt. **73**(2), 547–570 (2014). https://doi.org/10.1007/s10623-014-9953-5

Analysis of Reaction and Timing Attacks Against Cryptosystems Based on Sparse Parity-Check Codes

Paolo Santini[✉], Massimo Battaglioni, Franco Chiaraluce, and Marco Baldi

Università Politecnica delle Marche, Ancona, Italy
p.santini@pm.univpm.it, {m.battaglioni,f.chiaraluce,m.baldi}@univpm.it

Abstract. In this paper we study reaction and timing attacks against cryptosystems based on sparse parity-check codes, which encompass low-density parity-check (LDPC) codes and moderate-density parity-check (MDPC) codes. We show that the feasibility of these attacks is not strictly associated to the quasi-cyclic (QC) structure of the code but is related to the intrinsically probabilistic decoding of any sparse parity-check code. So, these attacks not only work against QC codes, but can be generalized to broader classes of codes. We provide a novel algorithm that, in the case of a QC code, allows recovering a larger amount of information than that retrievable through existing attacks and we use this algorithm to characterize new side-channel information leakages. We devise a theoretical model for the decoder that describes and justifies our results. Numerical simulations are provided that confirm the effectiveness of our approach.

Keywords: Code-based cryptosystems · LDPC codes · MDPC codes · Reaction attacks · Timing attacks

1 Introduction

The code-based cryptosystems introduced by McEliece [18] and Niederreiter [20] are among the oldest and most studied post-quantum public-key cryptosystems. They are commonly built upon a family of error correcting codes, for which an efficient decoding algorithm is known. The security of these systems is based on the hardness of the so called Syndrome Decoding Problem (SDP), i.e., the problem of decoding a random linear code, which has been proven to be NP-complete [7]. The best SDP solvers are known as Information Set Decoding (ISD) algorithms and were introduced in 1962 by Prange [23]; improved through the years (see [6,17,26], for some well known variants), these algorithms are characterized by exponential complexity, even when considering adversaries equipped with quantum computers [8]. Because of their well studied and assessed security,

---

The work of Paolo Santini was partially supported by Namirial S.p.A.

M. Baldi et al. (Eds.): CBC 2019, LNCS 11666, pp. 115–136, 2019.
https://doi.org/10.1007/978-3-030-25922-8_7

code-based cryptosystems are nowadays considered among the most promising candidates for the post-quantum world [9].

In the above schemes, and others of the same type, the private key is the representation of a code, whose parameters are chosen in such a way to guarantee decoding of a given amount of errors, which are intentionally introduced in the plaintext during encryption. The public key is obtained through linear transformations of the secret key, with the aim of masking the structure of the secret code. In the original McEliece proposal, Goppa codes were used: on the one hand, this choice leads to a well assessed security and the original proposal is still substantially unbroken; on the other hand, the corresponding public keys do not allow for any compact representation, and thus have very large sizes.

A well-known way to address the latter issue is that of adding some geometrical structure to the code, in order to guarantee that the public key admits a compact representation. The use of quasi-cyclic (QC) codes with a sparse parity-check matrix naturally fits this framework: the sparsity of the parity-check matrix allows for efficient decoding techniques, while the quasi-cyclic structure guarantees compactness in the public key. In such a context, the additional geometrical structure can be added without exposing the secret code: Quasi-Cyclic Low-Density Parity-Check (QC-LDPC) and Quasi-Cyclic Moderate-Density Parity-Check (QC-MDPC) code-based cryptosystems have been extensively studied in recent years [1,2,4,5,19] and achieve small public keys. However, differently from the bounded-distance decoders used for algebraic codes, like the mentioned Goppa codes, the iterative decoders used for sparse parity-check codes are not characterized by a deterministic decoding radius and, thus, decoding might fail with some probability, or Decoding Failure Rate (DFR).

Such a feature is crucial, since it has been shown how this probabilistic nature of the decoder actually exposes the system to cryptanalysis techniques based on the observation of the decryption phase. State-of-the-art attacks of this kind are commonly called reaction attacks, when based on decoding failure events [12,13,15,22], or side-channel attacks, when based on information such as the duration of the decoding phase, in which case we speak of timing attacks, or other quantities [10,11,21]. All these previous techniques exploit the QC structure of the code and aim at recovering some characteristics of the secret key by performing a statistical analysis on a sufficiently large amount of collected data. The rationale is that many quantities that are typical of the decryption procedure depend on a statistical relation between some properties of the secret key and the error vector that is used during encryption. Thus, after observing a sufficiently large number of decryption instances, an adversary can exploit the gathered information to reconstruct the secret key, or an equivalent version of it. The reconstruction phase is commonly very efficient, unless some specific choices in the system design are made [24,25] which, however, may have some significant drawbacks in terms of public key size. All the aforementioned attack techniques are instead prevented if the DFR has negligible values [27] and constant-time implementations of decryption algorithms are used. Alternatively, the use of ephemeral keys, which means that each key-pair is refreshed after just one decryption, is necessary to make these systems secure [1,5].

In this paper we study reaction and timing attacks, and we show that information leakage in the decoding phase can actually be related to the number of overlapping ones between columns of the secret parity-check matrix. Furthermore, we show that all attacks of this kind can be analyzed as a unique procedure, which can be applied to recover information about the secret key, regardless of the code structure. Such an algorithm, when applied on a QC code, allows recovering of an amount of information that is greater than that retrievable through previously published attacks. Moreover, we provide an approximated model that allows predicting the behaviour of the decoder in the first iteration with good accuracy. This model justifies the phenomenon that is at the basis of all the aforementioned attacks and can be even used to conjecture new attacks. Our results are confirmed by numerical simulations and enforce the employment of constant-time decoders with constant power consumption and negligible DFR, in order to allow for the use of long-lasting keys in these systems.

The paper is organized as follows: Sect. 2 describes the notation used throughout the manuscript and provides some basic notions about cryptosystems based on sparse parity-check codes. In Sect. 3 we summarize state-of-the-art reaction and timing attacks, and present a general algorithm that can be used to attack any sparse parity-check code. An approximate model for the analysis of the first iteration of the Bit Flipping (BF) decoder is presented in Sect. 4. In Sect. 5 we describe some additional sources of information leakage, which can be used by an opponent to mount a complete cryptanalysis of the system. Finally, in Sect. 6 we draw some final remarks.

2   Preliminaries

We represent matrices and vectors through bold capital and small letters, respectively. Given a matrix $\mathbf{A}$, we denote as $\mathbf{a}_i$ its i-th column and as $a_{i,j}$ its element in position (i, j). Given a vector $\mathbf{b}$, its i-th entry is referred to as b_i; its support is denoted as $\phi(\mathbf{b})$ and is defined as the set containing the positions of its non-null entries. The vector $\mathbf{0}_n$ corresponds to the all-zero n-tuple; the function returning the Hamming weight of its input is denoted as $\mathrm{wt}\{\cdot\}$.

2.1   LDPC and MDPC Code-Based Cryptosystems

The schemes we consider are built upon a code $\mathcal{C}$ described by a sparse parity-check matrix $\mathbf{H} \in \mathbb{F}_2^{r \times n}$, where n is the code blocklength. We here focus on the case of regular matrices, in which all the rows and all the columns have constant weights respectively equal to $w \ll n$ and $v \ll r$. The code $\mathcal{C}$ is then (v, w)-regular, and is commonly called Low-Density Parity-Check (LDPC) code if $w = O(\log n)$, or Moderate-Density Parity-Check (MDPC) code, if $w = O(\sqrt{n})$. Regardless of such a distinction, these two families of codes actually have similar properties: they can be decoded with the same decoding algorithms and are thus exposed in the same way to the attacks we consider. So, from now on we will not distinguish between these two families, and just refer to (v, w)-regular codes.

In the McEliece framework, the public key is a generator matrix $\mathbf{G}$ for $\mathcal{C}$; a ciphertext is obtained as

$$\mathbf{x} = \mathbf{mG} + \mathbf{e}, \tag{1}$$

where $\mathbf{m} \in \mathbb{F}_2^k$ is the plaintext and $\mathbf{e}$ is a randomly generated n-tuple with weight t. Decryption starts with the computation of the syndrome $\mathbf{s} = \mathbf{Hx}^T = \mathbf{He}^T$, where ^{T} denotes transposition. Then, an efficient syndrome decoding algorithm is applied on $\mathbf{s}$, in order to recover $\mathbf{e}$.

In the Niederreiter formulation, the public key is a parity-check matrix $\mathbf{H}' = \mathbf{SH}$ for $\mathcal{C}$, where $\mathbf{S}$ is a dense non singular matrix. The plaintext $\mathbf{m}$ is converted into an n-tuple $\mathbf{e}$ with weight t by means of an invertible mapping

$$\mathcal{M} \mapsto \{\mathbf{e} \in \mathbb{F}_2^n | \mathrm{wt}\{\mathbf{e}\} = t\}, \tag{2}$$

where $\mathcal{M}$ is the space of all possible plaintexts $\mathbf{m}$. The ciphertext is then computed as

$$\mathbf{x} = \mathbf{H}'\mathbf{e}^T. \tag{3}$$

Decryption starts with the computation of $\mathbf{s} = \mathbf{S}^{-1}\mathbf{x}$; then, an efficient syndrome decoding algorithm is applied on $\mathbf{s}$, in order to recover $\mathbf{e}$, from which the plaintext $\mathbf{m}$ is reconstructed by inverting (2).

Regardless of the particular system formulation we are considering (McEliece or Niederreiter), the decryption phase relies on a syndrome decoding algorithm, applied on the syndrome of a weight-t error vector. Since, in the attacks we consider, information is leaked during the decoding phase, we will not distinguish between the McEliece and the Niederreiter formulation in the following.

The decoding algorithm must show a good trade-off between complexity and DFR; for this reason, a common approach is that of relying on the so-called BF decoders, whose principle has been introduced by Gallager [14]. The description of a basic BF decoding procedure is given in Algorithm 1. The decoder goes through a maximum number of iterations $i_{\mathtt{max}}$, and at each iteration it exploits a likelihood criterion to estimate the error vector $\mathbf{e}$. Outputs of the decoder are the estimate of the error vector $\tilde{\mathbf{e}}$ and a boolean value $\perp$ reporting events of decoding failure. When $\perp = 0$, we have $\tilde{\mathbf{e}} = \mathbf{e}$, and decoding was successful; if $\perp = 1$, then $\tilde{\mathbf{e}} \neq \mathbf{e}$ and we have encountered a decoding failure. So, clearly, the DFR can be expressed as the probability that $\perp = 1$, noted as $P\{\perp = 1\}$. The likelihood criterion is based on a threshold b (line 11 of the algorithm), which, in principle, might also vary during the iterations (for instance, some possibilities are discussed in [19]); all the numerical results we show in this paper are referred to the simple case in which the threshold is kept constant throughout all the decoding procedure. In particular, in the simulations we have run, the values of $i_{\mathtt{max}}$ and b have been chosen empirically. Our analysis is general and can be easily extended to other decoders than those considered here. Indeed, many different decoders have been analyzed in the literature (for instance, see [10,21]), and, as for the outcome of reaction and timing attacks, there is no

Algorithm 1. BFdecoder

Input: $\mathbf{H} \in \mathbb{F}_2^{r \times n}$, $\mathbf{s} \in \mathbb{F}_2^r$, $i_{\max} \in \mathbb{N}$, $b \in \mathbb{N}$
Output: $\tilde{\mathbf{e}} \in \mathbb{F}_2^n$, $\bot \in \mathbb{F}_2$

1: $\tilde{\mathbf{e}} \leftarrow \mathbf{0}_n$
2: $\bot \leftarrow 0$
3: $i \leftarrow 0$
4: **while** $\mathrm{wt}\{\mathbf{s}\} > 0 \wedge i < i_{\max}$ **do**
5:      $\Psi \leftarrow \varnothing$
6:      **for** $j \leftarrow 0$ **to** $n - 1$ **do**
7:          $\sigma_j \leftarrow 0$
8:          **for** $l \in \phi(\mathbf{h}_j)$ **do**
9:              $\sigma_j \leftarrow \sigma_j + s_l$
10:          **end for**
11:          **if** $\sigma_j \geq b$ **then**
12:              $\Psi \leftarrow \Psi \cup j$ $\triangleright$ Position j is estimated as error affected
13:          **end if**
14:      **end for**
15:      **for** $j \in \Psi$ **do**
16:          $\tilde{e}_j \leftarrow \neg \tilde{e}_j$ $\triangleright$ Error estimation update
17:          $\mathbf{s} \leftarrow \mathbf{s} + \mathbf{h}_j$ $\triangleright$ Syndrome update
18:      **end for**
19:      $i \leftarrow i + 1$
20: **end while**
21: **if** $\mathrm{wt}\{\mathbf{s}\} > 0$ **then**
22:      $\bot \leftarrow 1$ $\triangleright$ Decoding failure
23: **end if**
24: **return** $\{\tilde{\mathbf{e}}, \bot\}$

meaningful difference between them. This strongly hints that such attacks are possible because of the probabilistic nature of the decoder, and are only slightly affected by the particular choice of the decoder and its settings. However, the analysis we provide in Sect. 4, which describes the decoder behaviour in the first iteration, takes into account the effect of the threshold value.

We point out that Algorithm 1 is commonly called an *out-of-place* decoder, as the syndrome $\mathbf{s}$ is updated after the set Ψ is computed. A different procedure is the one of *in-place* decoders, in which the syndrome is updated every time a bit is estimated as error affected (i.e., after the *if* instruction in line 11). In this paper we only focus on *out-of-place* decoders. The reason is that the attacks we consider seem to be emphasized when *in-place* decoders are used [10,21]. However, even if a careful analysis is needed, it is very likely that our results can be extended also to *in-place* decoders.

3   A General Framework for Reaction and Timing Attacks

In this section we describe a family of attacks based on statistical analyses, namely *statistical attacks*. This family includes reaction attacks, in which data is collected through the observation of Bob's reactions, and side-channel attacks. A statistical attack of the types here considered can be described as follows.

Let us consider a public-key cryptosystem with private and public keys K_S and K_P, respectively, and security parameter λ (i.e., the best attack on the system has complexity $> 2^\lambda$). We denote as $\mathsf{Decrypt}(K_S, \mathbf{x})$ a decryption algorithm that, given a ciphertext $\mathbf{x}$ and K_S as inputs, returns either the plaintext $\mathbf{m}$ or a decryption failure. We define $\mathcal{D}(K_S, \mathbf{x})$ as an oracle that, queried with a ciphertext $\mathbf{x}$, runs $\mathsf{Decrypt}(K_S, \mathbf{x})$ and returns some metrics that describe the execution of the decryption algorithm. More details about the oracle's replies are provided next. An adversary, which is given K_P, queries the oracle with N ciphertexts $\{\mathbf{x}^{(i)} \,|\, i = 1, \cdots, N\}$; we denote as $y^{(i)}$ the oracle's reply to the i-th query $\mathbf{x}^{(i)}$. The adversary then runs an algorithm $\mathcal{A}(K_P, \{\mathbf{x}^{(0)}, y^{(0)}\}, \cdots, \{\mathbf{x}^{(N-1)}, y^{(N-1)}\})$ that takes as inputs K_P and the pairs of oracle queries and replies, and returns K_S'. The algorithm $\mathcal{A}$ models the procedure that performs a statistical analysis of the gathered data and reconstructs the secret key, or an equivalent version of it. The time complexity of this whole procedure can be approximated as

$$C = \alpha N + C_{\mathcal{A}}, \tag{4}$$

where α corresponds to the average number of operations performed for each query and $C_{\mathcal{A}}$ is the complexity of executing the algorithm $\mathcal{A}$. The adversary is then challenged with a randomly generated ciphertext $\mathbf{x}^*$, corresponding to a plaintext $\mathbf{m}^*$. We consider the attack successful if $C < 2^\lambda$ and the probability of $\mathbf{m} = \mathsf{Decrypt}(K_S', \mathbf{x}^*)$ being equal to $\mathbf{m}^*$ is not negligible (i.e., larger than $2^{-\lambda}$).

We point out that this formulation is general, since it does not distinguish between the McEliece and Niederreiter cases. In the same way the private and public keys might be generic. For example, this model describes also reaction attacks against LEDA cryptosystems [5], in which the secret key consists of $\mathbf{H}$ and an additional sparse matrix $\mathbf{Q}$.

The above model allows for taking into account many kinds of attacks, depending on the oracle's reply. For instance, when considering attacks based on decryption failures, the oracle's reply is a boolean value which is true in case of a failure and false otherwise. When considering timing attacks based on the number of iterations, then the oracle's reply corresponds to the number of iterations run by the decoding algorithm.

In this paper we focus on systems with security against a Chosen Ciphertext Attack (CCA), that is, the case in which a proper conversion (like the one of [16]) is applied to the McEliece/Niederreiter cryptosystem, in order to achieve CCA security. In our attack model, this corresponds to assuming that the oracle queries are all randomly generated, i.e., the error vectors used during encryption can be seen as randomly picked elements from the ensemble of all n-uples with weight t. Opposed to the CCA case, in the Chosen Plaintext Attack (CPA) case the opponent is free to choose the error vectors used during encryption: from the adversary standpoint, the CPA assumption is clearly more optimistic than that of CCA, and leads to improvements in the attack [15,21]. Obviously, all results we discuss in this paper can be extended to the CPA case.

One final remark is about the schemes we consider: as shown in [24,25], the complexity of algorithm $\mathcal{A}$ can be increased with proper choices in the structure of the secret key. Basically, in these cases the adversary can gather information about the secret key, but cannot efficiently use this information to reconstruct the secret key, or to obtain an equivalent version of it. In this paper we do not consider such approaches and we assume that the algorithm $\mathcal{A}$ always runs in a feasible time, as it occurs in [12,15].

3.1 State-of-the-Art Statistical Attacks

Modern statistical attacks [10,12,13,15,21] are specific to the sole case of QC codes having the structure originally proposed in [3], which are defined through a secret parity-check matrix in the form

$$\mathbf{H} = \left[\mathbf{H}_0, \mathbf{H}_1, \ldots, \mathbf{H}_{n_0-1}\right], \tag{5}$$

where each $\mathbf{H}_i$ is a circulant matrix of weight v and size p, and n_0 is a small integer. Thus, the corresponding code is a (v, n_0v)-regular code.

All existing statistical attacks are focused on guessing the existence (or absence) of some cyclic distances between symbols 1 in $\mathbf{H}$. In particular, an adversary aims at recovering the following quantities, which were introduced in [15].

Distance Spectrum: *Given a vector* $\mathbf{a}$, *with support* $\phi(\mathbf{a})$ *and length* p, *its distance spectrum is defined as*

$$\mathrm{DS}(\mathbf{a}) = \left\{\min\{\pm(i-j) \mod p\}\,|\, i, j \in \phi(\mathbf{a}), \quad i \neq j\right\}. \tag{6}$$

Multiplicity: *We say that a distance* $d \in \mathrm{DS}(\mathbf{a})$ *has multiplicity* μ_d *if there are* μ_d *distinct pairs in* $\phi(\mathbf{a})$ *which produce the same distance* d.

Basically, the distance spectrum is the set of all distances with multiplicity larger than 0. It can be easily shown that all the rows of a circulant matrix are characterized by the same distance spectrum; thus, given a circulant matrix $\mathbf{M}$, we denote the distance spectrum of any of its rows (say, the first one) as $\mathrm{DS}(\mathbf{M})$.

Statistical attacks proposed in the literature aim at estimating the distance spectrum of the circulant blocks in the secret $\mathbf{H}$, and are based on the observation that some quantities that are typical of the decryption procedure depend on the number of common distances between the error vector and the rows of the parity-check matrix. In particular, the generic procedure of a statistical attack on a cryptosystem whose secret key is in the form (5) is described in Algorithm 2; we have called the algorithm Ex-GJS in order to emphasize the fact that it is an extended version of the original GJS attack [15], which was only focused on a single circulant block in $\mathbf{H}$. Our algorithm, which is inspired by that of [12], is a generalization of the procedure in [15], in which all the circulant blocks in $\mathbf{H}$ are taken into account. We present this algorithm in order to show the maximum amount of information that state-of-the-art statistical attacks allow to recover.

Algorithm 2. Ex-GJS

    Input: public key K_P, number of queries $N \in \mathbb{N}$
    Output: estimates $\mathbf{a}^{(0)}, \cdots, \mathbf{a}^{(n_0-1)}, \mathbf{b}^{(0)}, \cdots, \mathbf{b}^{(n_0-1)} \in \mathbb{N}_{\lfloor p/2 \rfloor}$.

1: Initialize $\mathbf{a}^{(0)}, \cdots, \mathbf{a}^{(n_0-1)}, \mathbf{b}^{(0)}, \cdots, \mathbf{b}^{(n_0-1)}$ as null arrays of length $\lfloor p/2 \rfloor$
2: **for** $i \leftarrow 1$ **to** N **do**
3:     $\mathbf{x}^{(i)} \leftarrow$ ciphertexts obtained through the error vector $\mathbf{e}^{(i)}$
4:     $y^{(i)} \leftarrow \mathcal{D}(K_S, \mathbf{x}^{(i)})$
5:     **for** $j \leftarrow 0$ **to** $n_0 - 1$ **do**
6:         $\Delta_j \leftarrow \mathrm{DS}(\mathbf{e}_j^{(i)})$
7:         **for** $d \in \Delta_j$ **do**
8:             $a_d^{(j)} \leftarrow a_d^{(j)} + y^{(i)}$
9:             $b_d^{(j)} \leftarrow b_d^{(j)} + 1$
10:         **end for**
11:     **end for**
12: **end for**
13: **return** $\{\mathbf{a}^{(0)}, \cdots, \mathbf{a}^{(n_0-1)}, \mathbf{b}^{(0)}, \cdots, \mathbf{b}^{(n_0-1)}\}$

The error vector used for the i-th query is expressed as $\mathbf{e}^{(i)} = [\mathbf{e}_0^{(i)}, \ldots, \mathbf{e}_{n_0-1}^{(i)}]$, where each $\mathbf{e}_j^{(i)}$ has length p. The estimates $\mathbf{a}^{(0)}, \ldots, \mathbf{a}^{(n_0-1)}$ and $\mathbf{b}^{(0)}, \ldots, \mathbf{b}^{(n_0-1)}$ are then used by the adversary to guess the distance spectra of the blocks in the secret key. Indeed, let us define $\mathcal{E}^{(d,j)}(n,t)$ as the ensemble of all error vectors having length n, weight t and such that they exhibit a distance d in the distance spectrum of the j-th length-p block. Then, depending on the meaning of the oracle's reply, the ratios $a_d^{(j)}/b_d^{(j)}$ correspond to the estimate of the average value of some quantity, when the error vector belongs to $\mathcal{E}^{(d,j)}(n,t)$. For instance, when considering attacks based on decryption failures, the oracle's reply is either 0 or 1, depending on whether the decryption was successful or failed. In such a case, the ratio $a_d^{(j)}/b_d^{(j)}$ corresponds to an empirical measurement of the DFR, conditioned to the event that the error vector belongs to $\mathcal{E}^{(j,d)}(n,t)$. In general, statistical attacks are successful because many quantities that are typical of the decoding procedure depend on the multiplicity of the distances in $\mathrm{DS}(\mathbf{H}_j)$, $\forall j \in [0, \ldots, n_0 - 1]$. In the next section we generalize this procedure, by considering different ensembles for the error vectors; then, in Sect. 4, we provide a theoretical explanation for such a phenomenon.

3.2 Exploiting Decryption Failures on Generic Codes

In this section we generalize the Ex-GJS procedure, and describe an algorithm which can be used to recover information about any regular code. In particular, our analysis shows that events of decoding failure (i) do not strictly depend on the QC structure of the adopted code, and (ii) permit to retrieve a quantity that is more general than distance spectra.

We first show that, for generic regular codes, there is a connection between the syndrome weight and the DFR. This statement is validated by numerical simulations on (v, w)-regular codes, obtained through Gallager construction [14], in which $\frac{v}{w} = \frac{r}{n}$. In particular, we have considered two codes with length $n = 5000$, redundancy $r = 2500$ and different pairs (v, w), decoded through Algorithm 1; their DFR (i.e., the probability of Algorithm 1 returning $\bot = 1$) vs. syndrome weight is shown in Fig. 1. We notice from Fig. 1 that there is a strong dependence between the initial syndrome weight and the DFR and that different pairs (v, w) can lead to two different trends in the DFR evolution. Section 4 is devoted to the explanation of this phenomenon.

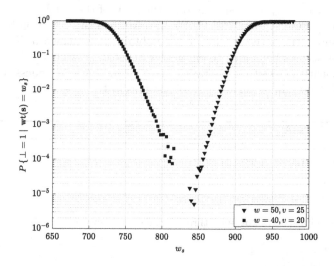

Fig. 1. Distribution of the DFR as a function of the syndrome weight, for two regular (v, w)-regular LDPC codes, decoded through Algorithm 1 with $i_{\max} = 5$ and $b = 15$. The weight of the error vectors is $t = 58$; for each code, 10^7 decoding instances have been considered.

Let us now define $\mathcal{E}(n, t, i_0, i_1)$ as the ensemble of all vectors having length n, weight t and whose support contains elements i_0 and i_1. Let $\mathbf{s}$ be the syndrome of an error vector $\mathbf{e} \in \mathcal{E}(n, t, i_0, i_1)$: we have

$$\mathbf{s} = \mathbf{h}_{i_0} + \mathbf{h}_{i_1} + \sum_{j \in \phi(\mathbf{e}) \backslash \{i_0, i_1\}} \mathbf{h}_j. \tag{7}$$

The syndrome weight has a probability distribution that depends on the interplay between $\mathbf{h}_{i_0}$ and $\mathbf{h}_{i_1}$: basically, when these two columns overlap in a small (large) number of ones, then the average syndrome weight gets larger (lower). Moreover, motivated by the empirical evidence of Fig. 1, one can expect that the DFR experienced over error vectors belonging to different ensembles $\mathcal{E}(n, t, i_0, i_1)$ depends on the number of overlapping ones between columns $\mathbf{h}_{i_0}$ and $\mathbf{h}_{i_1}$. Then, a statistical attack against a generic regular code can be mounted, as described in Algorithm 3, which we denote as General Statistical Attack (GSA). The output of the algorithm is represented by the matrices $\mathbf{A}$ and $\mathbf{B}$, which are used by the adversary to estimate the average value of the oracle's replies, as a function of the pair (i_0, i_1). Notice that in Algorithm 3 the oracle's reply is denoted as $y^{(i)}$ and does not need to be better specified. We will indeed show in Sect. 5 that the same procedure can be used to exploit other information sources than the success (or failure) of decryption. We now focus on the case of $y^{(i)}$ being 0 or 1, depending on whether decryption was successful or not. Then, each ratio $a_{j,l}/b_{j,l}$ represents an empirical estimate of the probability of encountering a decryption failure, when the error vector contains both j and l in its support. One might expect that the ratios $a_{j,l}/b_{j,l}$ are distributed on the basis of the number of overlapping ones between columns j and l in $\mathbf{H}$. We have verified this intuition by means of numerical simulations; the results we have obtained are shown in Fig. 2, for the case of error vectors belonging to ensembles $\mathcal{E}(n, t, 0, j)$, with $j \in [1, \ldots, n-1]$. The figure clearly shows that the ratios $a_{j,l}/b_{j,l}$ can be used to guess the number of overlapping ones between any pair of columns in $\mathbf{H}$.

Algorithm 3. GSA

    Input: public key K_P, number of queries $N \in \mathbb{N}$
    Output: estimates $\mathbf{A}, \mathbf{B} \in \mathbb{N}^{n \times n}$.

1: $\mathbf{A} \leftarrow \mathbf{0}_{n \times n}$
2: $\mathbf{B} \leftarrow \mathbf{0}_{n \times n}$
3: **for** $i \leftarrow 1$ **to** N **do**
4:     $\mathbf{x}^{(i)} \leftarrow$ ciphertexts obtained through the error vector $\mathbf{e}^{(i)}$
5:     $y^{(i)} \leftarrow \mathcal{D}(K_S, \mathbf{x}^{(i)})$
6:     $\phi(\mathbf{e}^{(i)}) \leftarrow$ support of $\mathbf{e}^{(i)}$
7:     **for** $j \in \phi(\mathbf{e}^{(i)})$ **do**
8:         **for** $l \in \phi(\mathbf{e}^{(i)})$ **do**
9:             $a_{j,l} \leftarrow a_{j,l} + y^{(i)}$
10:            $b_{j,l} \leftarrow b_{j,l} + 1$
11:         **end for**
12:     **end for**
13: **end for**
14: **return** $\{\mathbf{A}, \mathbf{B}\}$

These empirical results confirm the conjecture that the DFR corresponding to error vectors in $\mathcal{E}(n, t, i_0, i_1)$ depends on the number of overlapping ones between the columns i_0 and i_1. Moreover, these results show that the same idea of [15], with some generalization, can be applied to whichever kind of code.

We now show that even when QC codes are considered, our algorithm recovers more information than that which can be obtained through the Ex-GJS procedure. For such a purpose, let us consider a parity-check matrix in the form (5), and let $\gamma_{i,j}$ be the number of overlapping ones between columns i and j. Now, because of the QC structure, we have

$$\left| \phi(\mathbf{h}_i) \cap \phi(\mathbf{h}_j) \right| = \left| \phi(\mathbf{h}_{p\lfloor i/p \rfloor + [i+z \bmod p]}) \cap \phi(\mathbf{h}_{p\lfloor j/p \rfloor + [j+z \bmod p]}) \right|, \quad \forall z. \quad (8)$$

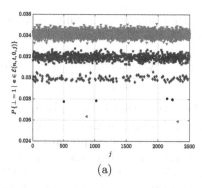 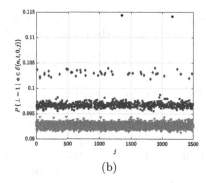

     (a)   (b)

Fig. 2. Simulation results for (v, w)-regular codes with $n = 5000$, $k = 2500$, for $t = 58$ and for error vector belonging to ensembles $\mathcal{E}(n, t, 0, j)$, for $j \in [1, \ldots, n-1]$. The parameters of the codes are $v = 25$, $w = 50$ for Figure (a), $v = 20$, $w = 40$ for Figure (b); the decoder settings are $i_{\max} = 5$ and $b = 15$. The results have been obtained through the simulation of 10^9 decoding instances. Grey, blue, green, black and red markers are referred to pairs of columns with number of intersections equal to $0, 1, 2, 3, 4$, respectively. (Color figure online)

We now consider two columns that belong to the same circulant block in $\mathbf{H}$, i.e. $i = pi_p + i'$, $j = pi_p + j'$, with $0 \leq i_p \leq n_0 - 1$ and $0 \leq i', j' \leq p - 1$; then, (8) can be rewritten as

$$|\phi(\mathbf{h}_i) \cap \phi(\mathbf{h}_j)| = |\phi(\mathbf{h}_{pi_p + [i' + z \bmod p]}) \cap \phi(\mathbf{h}_{pi_p + [j' + z \bmod p]})|, \quad \forall z. \quad (9)$$

With some simple computations, we finally obtain

$$|\phi(\mathbf{h}_{pi_p + i'}) \cap \phi(\mathbf{h}_{pi_p + j'})| = \begin{cases} |\phi(\mathbf{h}_{pi_p}) \cap \phi(\mathbf{h}_{pi_p + p - (i' - j')})| & \text{if } j' < i' \\ |\phi(\mathbf{h}_{pi_p}) \cap \phi(\mathbf{h}_{pi_p + j' - i'})| & \text{if } j' > i' \end{cases} \quad (10)$$

Similar considerations can be carried out if the two columns do not belong to the same circulant block. So, (10) shows that the whole information about overlapping ones between columns in $\mathbf{H}$ is actually represented by a subset of all the possible values of $\gamma_{i,j}$. This means that the execution of Algorithm 3 can be sped up by taking the QC structure into account.

In particular, the values of $\gamma_{i,j}$ can be used to obtain the distance spectra of the blocks in $\mathbf{H}$ in a straightforward way. Let us refer to Eq. (10), and look at two columns $\mathbf{h}_{pi_p}$ and $\mathbf{h}_j$, with $j = pi_p + j'$, where $j' \in [0, 1, \ldots, p - 1]$. The support of $\mathbf{h}_{pi_p}$ is $\phi(\mathbf{h}_{pi_p}) = \{c_0^{(pi_p)}, \cdots, c_{v-1}^{(pi_p)}\}$. The support of $\mathbf{h}_j$ can be expressed as

$$\phi(\mathbf{h}_j) = \{c_l^{(j)} \mid c_l^{(j)} = c_l^{(pi_p)} + j' \bmod p, \ l \in [0, \ldots, v-1], \ c_l^{(pi_p)} \in \phi(\mathbf{h}_{pi_p})\}. \quad (11)$$

Then, we have $|\phi(\mathbf{h}_{pi_p}) \cap \phi(\mathbf{h}_j)| = \gamma_{pi_p, j}$; this means that there are $\gamma_{pi_p, j}$ pairs $\{c, c'\} \in \phi(\mathbf{h}_{pi_p}) \times \phi(\mathbf{h}_j)$ such that

$$c' = c + d \bmod p, d \in \{j', p - j'\}. \quad (12)$$

It is easy to see that (12) corresponds to the definition of the distance spectrum of the blocks in $\mathbf{H}$; then, (12) can be turned into the following rule

$$\left|\phi(\mathbf{h}_{pi_p}) \cap \phi(\mathbf{h}_p i_p + j')\right| = \gamma \leftrightarrow d \in \mathrm{DS}(\mathbf{H}_{i_p}), \quad \mu_d = \gamma, \tag{13}$$

with $d = \min\{\pm j' \bmod p\}$.

This proves that the procedure described by Algorithm 3 allows obtaining at least the same amount of information recovered through the Ex-GJS algorithm, which is specific to the QC case and guarantees a complete cryptanalysis of the system [15]. In other words, our analysis confirms that Algorithm 3 is applicable and successful in at least all the scenarios in which the attack from [15] works. Moreover, our procedure allows for recovering a larger amount of information about the secret key, and thus defines a broader perimeter of information retrieval, which encompasses existing and future attacks.

4 An Approximate Model for Reaction Attacks

The main result of this section is summarized in the following proposition, for which we provide theoretical justifications and empirical evidences.

Proposition 1. *Let $\mathbf{H}$ be the parity-check matrix of a (v, w)-regular code, which is decoded through Algorithm 1 with decoding threshold b. Let (i_0, i_1) and (i_0^*, i_1^*) be two distinct pairs of indexes, and consider error vectors $\mathbf{e} \in \mathcal{E}(n, t, i_0, i_1)$ and $\mathbf{e}^* \in \mathcal{E}(n, t, i_0^*, i_1^*)$. Let ϵ and ϵ^* be the probabilities that $\mathbf{e}$ and $\mathbf{e}^*$ result in a decoding failure, respectively. Let $\mathbf{e}'_{[1]}$ be the error vector estimate after the first iteration; we define $\mathbf{e}' = \mathbf{e} + \mathbf{e}'_{[1]}$ and $t' = \mathrm{wt}\{\mathbf{e}'\}$. Then, $\epsilon > \epsilon^*$ if and only if $E[t'] > E[t'^*]$, where $E[\cdot]$ denotes the expected value.*

Essentially, the above proposition implies that increments (reductions) of the DFR are due to the fact that, depending on the particular matrix $\mathbf{H}$, some error patterns tend to produce, on average, a larger (lower) amount of residual errors, after the first decoder iteration. First of all, this statement is actually supported by empirical evidences: we have run numerical simulations on the same codes as those in Figs. 1 and 2, and have evaluated the number of residual errors after the first iteration. The results are shown in Fig. 3; as we can see, accordingly with Proposition 1, the trend of the DFR and the one of t' are the same for the analyzed codes.

We now derive a statistical model which approximates how the BF decoder described in Algorithm 1 evolves during the first iteration; through this model we can predict the value of t' and, thus, also justify the different trends of the DFR observed in Figs. 1 and 2. We choose two distinct integers i_0, i_1 and consider the case of an error vector randomly drawn from the ensemble $\mathcal{E}(n, t, i_0, i_1)$, that is, we take columns i_0 and i_1 of $\mathbf{H}$ as a reference, assuming that $e_{i_0} = e_{i_1} = 1$. We also suppose that the columns i_0 and i_1 of $\mathbf{H}$ overlap in γ positions, and aim at expressing the average value of t' as a function of the code parameters, the decoding threshold b and the value of γ.

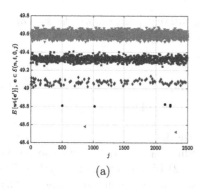 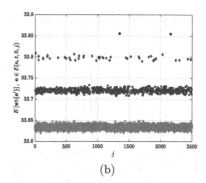

(a)   (b)

Fig. 3. Simulation results for the number of residual errors after the first iteration of BF decoding, for (v, w)-regular codes with $n = 5000$, $k = 2500$, for $t = 58$ and for error vector drawn from ensembles $\mathcal{E}(n, t, 0, j)$, for $j \in [1, \ldots, n-1]$. The parameters of the codes are $v = 25$, $w = 50$ for Figure (a), $v = 20$, $w = 40$ for (b); the decoder settings are $i_{\max} = 5$ and $b = 15$. The results have been obtained through the simulation of 10^9 decoding instances. Grey, blue, green, black and red markers are referred to pairs of columns with number of intersections equal to $0, 1, 2, 3, 4$, respectively. (Color figure online)

Let us first partition the sets of parity-check equations and variable nodes as follows.

Definition 1. Given an $r \times n$ parity-check matrix $\mathbf{H}$, the set $\{0, 1, \cdots, r-1\}$ can be partitioned into three subsets, defined as follows:

(i) $\mathcal{P}_0$: the set of parity-check equations that involve both bits i_0 and i_1, that is

$$j \in \mathcal{P}_0 \text{ iff } h_{j,i_0} = 1 \wedge h_{j,i_1} = 1;$$

(ii) $\mathcal{P}_1$: the set of parity-check equations that involve either bit i_0 or bit i_1, that is

$$j \in \mathcal{P}_1 \text{ iff } (h_{j,i_0} = 1 \wedge h_{j,i_1} = 0) \vee (h_{j,i_0} = 0 \wedge h_{j,i_1} = 1);$$

(iii) $\mathcal{P}_2$: the set of parity-check equations that do not involve bits i_0 and i_1, that is

$$j \in \mathcal{P}_2 \text{ iff } h_{j,i_0} = 0 \wedge h_{j,i_1} = 0.$$

Definition 2. Let $\mathcal{V}_i$, with $i \in \{0, 1, 2\}$, be the set defined as

$$\mathcal{V}_i = \begin{cases} \{l \in [0; n-1] \setminus \{i_0, i_1\} | \, \exists j \in \mathcal{P}_i \text{ s.t. } h_{j,l} = 1\} & \text{if } i = 0, 1 \\ \{l \in [0; n-1] \setminus \{i_0, i_1\} | \, \nexists j \in \mathcal{P}_0 \cup \mathcal{P}_1 \text{ s.t. } h_{j,l} = 1\} & \text{if } i = 2 \end{cases}. \quad (14)$$

The cardinality of each set $\mathcal{V}_i$ depends on the particular matrix $\mathbf{H}$. However, when considering a regular code, we can derive some general properties, as stated in the following lemma.

Lemma 1. *Given a (v, w)-regular code, the following bounds on the size of the sets $\mathcal{V}_i$, with $i \in \{0, 1, 2\}$, as defined in Definition 2, hold*

$$\begin{cases} |\mathcal{V}_0| \leq \gamma(w - 2), \\ |\mathcal{V}_1| \leq (2v - 2\gamma)(w - 1), \\ |\mathcal{V}_2| \geq n - 2 + \gamma w - 2v(w - 1). \end{cases} \tag{15}$$

Proof. The first bound in (15) follows from the fact that $|\mathcal{P}_0| = \gamma$. Any parity-check equation in $\mathcal{P}_0$ involves w bits, including i_0 and i_1. So, all the parity-check equations in $\mathcal{P}_0$ involve, at most, $\gamma(w - 2)$ bits other than i_0 and i_1. The second bound in (15) can be derived with similar arguments, considering that $|\mathcal{P}_1| = 2(v - \gamma)$, and either i_0 or i_1 participates to the parity-check equations in $\mathcal{P}_1$. The third bound is simply obtained by considering the remaining $r + \gamma - 2v$ parity-check equations, when the other two bounds in (15) hold with equality sign.

From now on, in order to make our analysis as general as possible, i.e., independent of the particular $\mathbf{H}$, we make the following assumption.

Assumption 1. *Let $\mathbf{H}$ be the $r \times n$ parity-check matrix of a (v, w)-regular code. We assume that each variable node other than i_0 and i_1 either participates in just one parity-check equation from $\mathcal{P}_0$ or $\mathcal{P}_1$ and $v - 1$ equations in $\mathcal{P}_2$, or participates in v parity-check equations in $\mathcal{P}_2$. This means that*

$$\begin{cases} |\mathcal{V}_0| = \gamma(w - 2), \\ |\mathcal{V}_1| = (2v - 2\gamma)(w - 1), \\ |\mathcal{V}_2| = n - 2 + \gamma w - 2v(w - 1). \end{cases} \tag{16}$$

The previous assumption is justified by the fact that, due to the sparsity of $\mathbf{H}$, we have $r \gg v$; it then follows that $|\mathcal{P}_0| \ll |\mathcal{P}_2|$ and $|\mathcal{P}_1| \ll |\mathcal{P}_2|$. Clearly, this assumption becomes more realistic as the matrix $\mathbf{H}$ gets sparser.

We additionally define $t^{(i)} = |\phi(\mathbf{e}) \cap \mathcal{V}_i|$; clearly, we have $t^{(0)} + t^{(1)} + t^{(2)} = t - 2$. The probability of having a specific configuration of $t^{(0)}$, $t^{(1)}$ and $t^{(2)}$ is equal to

$$P_{\{t^{(0)}, t^{(1)}, t^{(2)}\}} = \frac{\binom{|\mathcal{V}_0|}{t^{(0)}} \binom{|\mathcal{V}_1|}{t^{(1)}} \binom{|\mathcal{V}_2|}{t^{(2)}}}{\binom{n-2}{t-2}}. \tag{17}$$

In analogous way, we define $t^{(i)'}$ as the number of nodes that are in $\mathcal{V}^{(i)}$ and are simultaneously set in $\mathbf{e} + \mathbf{e}'_{[1]}$. In other words, $t^{(i)'}$ corresponds to the number of errors that, after the first iteration, affect bits in $\mathcal{V}_i$.

The definitions of the sets $\mathcal{V}_i$ are useful to analyze how the value of γ influences the decoder choices. We focus on a generic j-th bit, with $j \neq i_0, i_1$, and consider the value of σ_j, as defined in Algorithm 1. Because of Assumption 1, we have that

1. if $j \in \mathcal{V}_0$ (resp. $\mathcal{V}_1$), the j-th bit participates in one parity-check equation from $\mathcal{P}_0$ (resp. $\mathcal{V}_1$) and $v - 1$ parity-check equations in $\mathcal{P}_2$;
2. if $j \in \mathcal{V}_2$, the j-th bit participates in v parity-check equations in $\mathcal{P}_2$;
3. if $j \in \{i_0, i_1\}$, then it participates in γ parity-check equations from $\mathcal{P}_0$ and $v - \gamma$ parity-check equations from $\mathcal{P}_1$.

Let $p_{d,u}^{(i)}$, with $i = \{0, 1, 2\}$, be the probability that a parity-check equation involving the j-th bit (with $j \neq i_0, i_1$) and contained in $\mathcal{P}_i$ is unsatisfied, in the case of $e_j = d$, with $d \in \{0, 1\}$; the value of such a probability is expressed by Lemma 2.

Lemma 2. *Let us consider a (v, w)-regular code with blocklength n and an error vector $\mathbf{e}$ with weight t. Then, the probabilities $p_{d,u}^{(i)}$, with $i \in \{0, 1, 2\}$ and $d \in \{0, 1\}$, can be calculated as*

$$p_{0,u}^{(0)} = \sum_{\substack{l=1 \\ l \ odd}}^{\min\{w-3, t-2\}} \frac{\binom{w-3}{l}\binom{n-w}{t-l-2}}{\binom{n-3}{t-2}},$$

$$p_{0,u}^{(1)} = \sum_{\substack{l=0 \\ l \ even}}^{\min\{w-2, t-2\}} \frac{\binom{w-2}{l}\binom{n-w-1}{t-l-2}}{\binom{n-3}{t-2}},$$

$$p_{1,u}^{(0)} = \sum_{\substack{l=0 \\ l \ even}}^{\min\{w-3, t-3\}} \frac{\binom{w-3}{l}\binom{n-w}{t-l-3}}{\binom{n-3}{t-3}},$$

$$p_{1,u}^{(1)} = \sum_{\substack{l=1 \\ l \ odd}}^{\min\{w-2, t-3\}} \frac{\binom{w-2}{l}\binom{n-w-1}{t-l-3}}{\binom{n-3}{t-3}},\tag{18}$$

$$p_{0,u}^{(2)} = \sum_{\substack{l=1 \\ l \ odd}}^{\min\{w-1, t-2\}} \frac{\binom{w-1}{l}\binom{n-w-2}{t-l-2}}{\binom{n-3}{t-2}},$$

$$p_{1,u}^{(2)} = \sum_{\substack{l=0 \\ l \ even}}^{\min\{w-1, t-3\}} \frac{\binom{w-1}{l}\binom{n-w-2}{t-l-3}}{\binom{n-3}{t-3}}.$$

Proof. Let us first consider $i = 0$ and $d = 0$. Let us also consider the j-th bit, different from i_0 and i_1. Any parity check equation in $\mathcal{P}_0$ overlaps with the error vector in two positions, as it involves both bits i_0 and i_1; since we are looking at an error-free bit, then the parity-check equation will be unsatisfied only if the remaining $t - 2$ errors intercept an odd number of ones, among the remaining $w - 3$ ones. Simple combinatorial arguments lead to the first expression of (18). All the other expressions can be derived with similar arguments. $\qquad\square$

We also define $p_{\mathcal{E},u}^{(i)}$, with $i \in \{0,1\}$, as the probability that a parity-check equation involving a bit $\in \{i_0, i_1\}$, and contained in $\mathcal{P}_i$, is unsatisfied; the value of such a probability is derived in Lemma 3.

Lemma 3. *Let us consider a (v,w)-regular code with blocklength n and an error vector $\mathbf{e}$ with weight t. Then, the probabilities $p_{\mathcal{E},u}^{(i)}$, with $i \in \{0,1\}$, can be calculated as*

$$p_{\mathcal{E},u}^{(0)} = \sum_{\substack{l=1 \\ l \text{ odd}}}^{\min\{w-2,t-2\}} \frac{\binom{w-2}{l}\binom{n-w}{t-2-l}}{\binom{n-2}{t-2}}, \tag{19}$$

$$p_{\mathcal{E},u}^{(1)} = \sum_{\substack{l=0 \\ l \text{ even}}}^{\min\{w-1,t-2\}} \frac{\binom{w-1}{l}\binom{n-w-1}{t-2-l}}{\binom{n-2}{t-2}}. \tag{20}$$

Proof. The proof can be carried on with the same arguments of the proof of Lemma 2. □

We now consider the following assumption.

Assumption 2. *Let $\mathbf{H}$ be the parity-check matrix of a (v,w)-regular code. We assume that the parity-check equations in which the j-th bit is involved are statistically independent; thus, σ_j, defined as in Algorithm 1, can be described in the first decoding iteration as the sum of independent Bernoulli random variables, each one having its own probability of being set, which corresponds either to $p_{d,u}^{(i)}$ or $p_{\mathcal{E},u}^{(d)}$, where $i \in \{0,1,2\}$ and $d \in \{0,1\}$.*

We now define $P_{d,\text{flip}}^{(i)}$ as the probability that the decoder flips the j-th bit, in the case that $j \neq i_0, i_1$ and $j \in \mathcal{V}_i$, when $e_j = d$. In analogous way, $P_{\mathcal{E},\text{flip}}$ denotes the probability that, when $j \in \{i_0, i_1\}$, the decoder flips the j-th bit. The above probabilities are computed in Lemmas 4 and 5, respectively.

Lemma 4. *Let us consider a (v,w)-regular code with blocklength n and an error vector $\mathbf{e}$ with weight t; let b denote the decoding threshold employed in the first iteration. Then, under Assumptions 1 and 2, the probabilities $P_{d,\text{flip}}^{(i)}$, with $i \in \{0,1,2\}$ and $d \in \{0,1\}$, can be computed as follows*

$$P_{d,\text{flip}}^{(0)} = P\{\sigma_j^{(2)} = b-1|e_j = d\}p_{d,u}^{(0)} + \sum_{l=b}^{v-1} P\{\sigma_j^{(2)} = l|e_j = d\}, \tag{21}$$

$$P_{d,\text{flip}}^{(1)} = P\{\sigma_j^{(2)} = b-1|e_j = d\}p_{d,u}^{(1)} + \sum_{l=b}^{v-1} P\{\sigma_j^{(2)} = l|e_j = d\}. \tag{22}$$

$$P_{d,\text{flip}}^{(2)} = \sum_{l=b}^{v} \binom{v}{l} \left(p_{d,u}^{(2)}\right)^l \left(1 - p_{d,u}^{(2)}\right)^{v-l}. \tag{23}$$

Proof. When $j \in \mathcal{V}_0$, the j-th bit is involved in one parity-check equation in $\mathcal{P}_0$ and $v - 1$ equations in $\mathcal{P}_2$. The probability that the decoder in the first iteration flips the j-th bit can be computed as

$$P_{d,\text{flip}}^{(0)} = P\{\sigma_j^{(2)} = b - 1|e_j = d\}p_{d,u}^{(0)} + P\{\sigma_j^{(2)} \geq b|e_j = d\}. \tag{24}$$

In particular, we have

$$P\{\sigma_j^{(2)} = z|e_j = d\} = \binom{v-1}{z} \left(p_{d,u}^{(2)}\right)^z \left(1 - p_{d,u}^{(2)}\right)^{v-1-z}, \tag{25}$$

so that

$$P_{d,\text{flip}}^{(0)} = P\{\sigma_j^{(2)} = b - 1|e_j = d\}p_{d,u}^{(0)} + \sum_{l=b}^{v-1} P\{\sigma_j^{(2)} = l|e_j = d\}. \tag{26}$$

Similarly, if $j \in \mathcal{V}_1$, then it is involved in one parity-check equation in $\mathcal{P}_1$ and $v - 1$ equations in $\mathcal{P}_2$; thus, we have

$$P_{d,\text{flip}}^{(1)} = P\{\sigma_j^{(2)} = b - 1|e_j = d\}p_{d,u}^{(1)} + \sum_{l=b}^{v-1} P\{\sigma_j^{(2)} = l|e_j = d\}. \tag{27}$$

Finally, if $j \in \mathcal{V}_2$, then it is involved in v parity-check equations in $\mathcal{V}_2$; using a similar reasoning as in the previous cases, we can write

$$P_{d,\text{flip}}^{(2)} = \sum_{l=b}^{v} \binom{v}{l} \left(p_{d,u}^{(2)}\right)^l \left(1 - p_{d,u}^{(2)}\right)^{v-l}. \tag{28}$$

This proves the lemma.   □

Lemma 5. *Let us consider a (v, w)-regular code with blocklength n and an error vector $\mathbf{e}$ with weight t; let b denote the decoding threshold employed in the first iteration. Then, under Assumptions 1 and 2, the probability $P_{\mathcal{E},\text{flip}}$ can be computed as follows*

$$P_{\mathcal{E},\text{flip}} = \sum_{l^{(0)}=0}^{\gamma} \sum_{l^{(1)}=b-l^{(0)}}^{v-\gamma} P\{\sigma_{\mathcal{E}}^{(0)} = l^{(0)}\}P\{\sigma_{\mathcal{E}}^{(1)} = l^{(1)}\}, \tag{29}$$

where

$$\begin{cases} P\{\sigma_{\mathcal{E}}^{(0)} = l\} = \binom{\gamma}{l} \left(p_{\mathcal{E},u}^{(0)}\right)^l \left(1 - p_{\mathcal{E},u}^{(0)}\right)^{\gamma-l}, \\ P\{\sigma_{\mathcal{E}}^{(1)} = l\} = \binom{v-\gamma}{l} \left(p_{\mathcal{E},u}^{(1)}\right)^l \left(1 - p_{\mathcal{E},u}^{(1)}\right)^{v-\gamma-l}. \end{cases} \tag{30}$$

Proof. Equation (30) derives from the fact that bits i_0 and i_1 participate in γ parity-check equations in $\mathcal{P}_0$; furthermore, both i_0 and i_1 participate in $v - \gamma$ equations in $\mathcal{P}_1$ each. Then, (29) expresses the probability that the number of unsatisfied parity-check equations for bit i_0 or i_1 is not smaller than the threshold b.   □

In order to estimate the average number of bits flipped after one iteration, we have to consider all the possible configurations of the error vector $\mathbf{e}$. The average value of $t^{(i)'}$ for the bits which are not in $\mathcal{E}$ can be computed as

$$E\left[t^{(i)'}\right] = t^{(i)}\left(1 - P^{(i)}_{1,\text{flip}}\right) + \left(|\mathcal{V}_i| - t^{(i)}\right)P^{(i)}_{0,\text{flip}}, \tag{31}$$

and the average number of errors in all bits $\mathcal{V} = \bigcup_{i=0}^{2}\mathcal{V}_i$ can be estimated as

$$E\left[t'_{\mathcal{V}}\right] = \sum_{t^{(0)}=0}^{\min\{t-2,|\mathcal{V}_0|\}}\sum_{t^{(1)}=0}^{\min\{t^{(1)}+t^{(2)},|\mathcal{V}_1|\}} E_{\{t^{(0)},t^{(1)},t^{(2)}\}} \cdot P_{\{t^{(0)},t^{(1)},t^{(2)}\}}, \tag{32}$$

where $t^{(2)} = t-2-(t^{(0)}+t^{(1)})$ and $E_{\{t^{(0)},t^{(1)},t^{(2)}\}} = E\left[t^{(0)'}\right]+E\left[t^{(1)'}\right]+E\left[t^{(2)'}\right]$.

Similarly, the average number of residual errors due to the bits in $\mathcal{E}$ can be derived as

$$E[t'_{\mathcal{E}}] = 2(1 - P_{\mathcal{E},\text{flip}}). \tag{33}$$

We can finally obtain the average value of t' over all bits in $\{0, 1, \ldots, n-1\}$ as

$$E\left[t'\right] = E[t'_{\mathcal{E}}] + E\left[t'_{\mathcal{V}}\right]. \tag{34}$$

A comparison between the simulated average values of t' and the theoretical ones is shown in Table 1 for the two codes already considered in Figs. 2 and 3, as a function of γ. As we can see, this model allows for a close prediction of the average value of t' starting from the number of overlapping ones γ. This also allows an accurate modeling of the behaviour of the number of errors (increasing or decreasing) as a function of γ.

Table 1. Average values of t', for different values of γ. Code (a) and Code (b) are the same as those considered in Figs. 2 and 3.

γ	Code (a)		Code (b)	
	Simulated	Theoretical	Simulated	Theoretical
0	49.59	48.98	33.63	33.64
1	49.33	48.76	33.72	33.73
2	49.07	48.55	33.80	33.80
3	48.81	48.34	33.86	33.86
4	48.55	48.15	-	-

5   Other Sources of Information Leakage

In this section we show some additional information leaks that might be exploited by an adversary to gather information about the structure of the secret $\mathbf{H}$.

The results in the previous section show how, on average, the number of residual errors after the first iteration can be associated to the number of over-lapping ones between columns of $\mathbf{H}$. Then, if the opponent has access to $\mathbf{e}'_{[1]}$ (i.e., to the positions that have been flipped in the first iteration), he can suc-ceed in recovering the values of $\gamma_{i,j}$. Indeed, once $\mathbf{e}'_{[1]}$ is known, the opponent can compute the number of residual errors for each query as $\mathbf{e} + \mathbf{e}'_{[1]}$. Basically, this statistical attack can be modeled through Algorithm 3, by assuming that the oracle's answer $y^{(i)}$ is t', that is, the weight of $\mathbf{e} + \mathbf{e}'_{[1]}$. The results in the previous section clearly show that this procedure allows for the cryptanalysis of the system.

We point out that, in a practical scenario, the locations of the bits that have been flipped by the decoder can be estimated through some power analysis attack, as in [11]. This information might be masked through proper implemen-tation strategies; for instance, random permutations might be applied to the order of processing bits in the decoder. This solution, which was proposed by the authors of [11] as a countermeasure to the attack they introduced in the same paper, is however likely not to be strong enough for guaranteeing prevention of other kinds of information leaks.

For instance, let us suppose that the oracle's reply in Algorithm 3 is the weight of $\mathbf{e}'_{[1]}$, i.e., to the number of flips performed in the first iteration. In a real case scenario, estimating this quantity might not be too hard. Indeed, each flip requires the update of the error vector (one operation) and the update of the syndrome (v operations). Thus, we might expect that the duration of the first iteration, and/or its power consumption, linearly increases with the weight of $\mathbf{e}'_{[1]}$. It can be shown that also this quantity depends on the number of intersections between columns.

Indeed, let us recall the notation adopted in the previous section, and define as $N^{(i)}_{\mathcal{V},\mathrm{flip}}$ the average number of flips performed among nodes in $\mathcal{V}_i$. We can write

$$N^{(i)}_{\mathcal{V},\mathrm{flip}} = t^{(i)'} P^{(i)}_{1,\mathrm{flip}} + (|\mathcal{V}_i| - t^{(i)'}) P^{(i)}_{0,\mathrm{flip}}, \tag{35}$$

so that the average number of flips in $\mathcal{V}$ is

$$N_{\mathcal{V},\mathrm{flip}} = \sum_{t^{(0)}=0}^{\min\{t-2,|\mathcal{V}_0|\}} \sum_{t^{(1)}=0}^{\min\{t^{(1)}+t^{(2)},|\mathcal{V}_1|\}} P_{\{t^{(0)},t^{(1)},t^{(2)}\}} \left[N^{(0)}_{\mathcal{V},\mathrm{flip}} + N^{(1)}_{\mathcal{V},\mathrm{flip}} + N^{(2)}_{\mathcal{V},\mathrm{flip}} \right]. \tag{36}$$

where $t^{(2)} = t - 2 - (t^{(0)} + t^{(1)})$.

The average number of flips for the bits in $\mathcal{E}$ is equal to $N_{\mathcal{E},\mathrm{flip}} = 2P_{\mathcal{E},\mathrm{flip}}$. So, combining the effect of the above equations, we have

$$N_{\mathrm{flip}} = N_{\mathcal{E},\mathrm{flip}} + N_{\mathcal{V},\mathrm{flip}}. \tag{37}$$

The probabilities in (36) depend on the value of γ; so, statistical attacks based on this quantity are expected to be successful. We have verified this intuition by means of numerical simulations, and the results are shown at the end of this section.

Another quantity that might leak information about the secret key is represented by the evolution of the syndrome weight during iterations. Authors in [10] have shown that the weight of the initial syndrome $\mathbf{s} = \mathbf{He}^T$ reveals information about the secret key; making a little step forward, we show that this consideration is indeed general and holds also for the first iteration. We model this attack by assuming that an adversary runs Algorithm 3 and the oracle replies with the syndrome weight after the first iteration, i.e. the weight of $\mathbf{s}' = \mathbf{H}(\mathbf{e} + \mathbf{e}'_{[1]})^T$. In general, we expect $t' \ll n$. On the other hand, large values of t' are associated to large weights of $\mathbf{s}'$ as well. Since we have verified in the previous sections that error vectors drawn from ensembles $\mathcal{E}(n, t, i_0, i_1)$ are associated to different values of t', it follows that also the syndrome weight depends on the number of intersections between columns of $\mathbf{H}$.

We have verified all these ideas by means of numerical simulations. In particular, we have considered QC codes, described by $\mathbf{H}$ in the form (5), in the case of $n_0 = 2$, $p = 4801$ and $v = 45$.

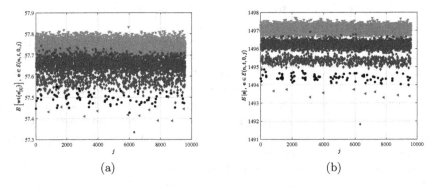

(a)   (b)

Fig. 4. Simulation results for statistical attacks based on t' and wt$\{\mathbf{s}'\}$. (a) shows the distribution of the values of t', (b) shows the weight of s'. The decoding threshold is $b = 28$; results have been obtained through the simulation of 10^8 decoding instances. Grey, blue, green, black, red and violet markers are referred to pairs of columns with number of intersections equal to $0, 1, 2, 3, 4, 5$, respectively. (Color figure online)

We have chosen $t = 84$ and $b = 28$, and applied Algorithm 3 with $N = 10^8$, considering that the oracle replies with: (i) the average number of errors after the first iteration in Fig. 4(a), (ii) the average weight of the syndrome vector after the first iteration in Fig. 4(b). This empirical evidence confirms our conjectures, and proves that these data, when leaked, might lead to a complete cryptanalysis of the system.

6   Conclusion

In this paper we have provided a deep analysis of statistical attacks against LDPC and MDPC code-based cryptosystems. We have considered a simple BF decoder, and have shown that its probabilistic nature might yield a substantial information leakage. We have shown that a general model for statistical attacks can be defined, and that many quantities, when observed by the opponent, might be exploited to recover information about the secret key. Our analysis confirms that, in order to safely use long-lasting keys in McEliece cryptosystem variants based on sparse parity-check matrix codes, a constant-time and constant-power implementation is necessary, along with a negligible DFR.

References

1. Aragon, N., et al.: BIKE: bit flipping key encapsulation (2017). http://bikesuite. org/files/BIKE.pdf
2. Baldi, M., Bodrato, M., Chiaraluce, F.: A new analysis of the McEliece cryptosystem based on QC-LDPC codes. In: Ostrovsky, R., De Prisco, R., Visconti, I. (eds.) SCN 2008. LNCS, vol. 5229, pp. 246–262. Springer, Heidelberg (2008). https:// doi.org/10.1007/978-3-540-85855-3_17
3. Baldi, M., Chiaraluce, F.: Cryptanalysis of a new instance of McEliece cryptosystem based on QC-LDPC codes. In: IEEE International Symposium on Information Theory (ISIT 2007), pp. 2591–2595, June 2007
4. Baldi, M., Barenghi, A., Chiaraluce, F., Pelosi, G., Santini, P.: LEDAkem: Low dEnsity coDe-bAsed key encapsulation mechanism (2017). https://www.ledacrypt. org/
5. Baldi, M., Barenghi, A., Chiaraluce, F., Pelosi, G., Santini, P.: LEDAkem: a post-quantum key encapsulation mechanism based on QC-LDPC codes. In: Lange, T., Steinwandt, R. (eds.) PQCrypto 2018. LNCS, vol. 10786, pp. 3–24. Springer, Cham (2018). https://doi.org/10.1007/978-3-319-79063-3_1
6. Becker, A., Joux, A., May, A., Meurer, A.: Decoding random binary linear codes in $2^{n/20}$: how $1 + 1 = 0$ improves information set decoding. In: Pointcheval, D., Johansson, T. (eds.) EUROCRYPT 2012. LNCS, vol. 7237, pp. 520–536. Springer, Heidelberg (2012). https://doi.org/10.1007/978-3-642-29011-4_31
7. Berlekamp, E., McEliece, R.J., van Tilborg, H.: On the inherent intractability of certain coding problems. IEEE Trans. Inf. Theory **24**(3), 384–386 (1978)
8. Bernstein, D.J.: Grover vs. McEliece. In: Sendrier, N. (ed.) PQCrypto 2010. LNCS, vol. 6061, pp. 73–80. Springer, Heidelberg (2010). https://doi.org/10.1007/978-3-642-12929-2_6
9. Chen, L., et al.: Report on post-quantum cryptography. Technical report NISTIR 8105, National Institute of Standards and Technology (2016)
10. Eaton, E., Lequesne, M., Parent, A., Sendrier, N.: QC-MDPC: a timing attack and a CCA2 KEM. In: Lange, T., Steinwandt, R. (eds.) PQCrypto 2018. LNCS, vol. 10786, pp. 47–76. Springer, Cham (2018). https://doi.org/10.1007/978-3-319-79063-3_3
11. Fabšič, T., Gallo, O., Hromada, V.: Simple power analysis attack on the QC-LDPC McEliece cryptosystem. Tatra Mt. Math. Pub. **67**(1), 85–92 (2016)
12. Fabšič, T., Hromada, V., Stankovski, P., Zajac, P., Guo, Q., Johansson, T.: A reaction attack on the QC-LDPC McEliece cryptosystem. In: Lange, T., Takagi, T. (eds.) PQCrypto 2017. LNCS, vol. 10346, pp. 51–68. Springer, Cham (2017). https://doi.org/10.1007/978-3-319-59879-6_4
13. Fabšič, T., Hromada, V., Zajac, P.: A reaction attack on LEDApkc. IACR Cryptology ePrint Archive 2018, 140 (2018)

14. Gallager, R.G.: Low-Density Parity-Check Codes. MIT Press, Cambridge (1963)
15. Guo, Q., Johansson, T., Stankovski, P.: A key recovery attack on MDPC with CCA security using decoding errors. In: Cheon, J.H., Takagi, T. (eds.) ASIACRYPT 2016. LNCS, vol. 10031, pp. 789–815. Springer, Heidelberg (2016). https://doi.org/10.1007/978-3-662-53887-6_29
16. Kobara, K., Imai, H.: Semantically secure McEliece public-key cryptosystems – conversions for McEliece PKC. In: Kim, K. (ed.) PKC 2001. LNCS, vol. 1992, pp. 19–35. Springer, Heidelberg (2001). https://doi.org/10.1007/3-540-44586-2_2
17. Lee, P.J., Brickell, E.F.: An observation on the security of McEliece's public-key cryptosystem. In: Barstow, D., Brauer, W., Brinch Hansen, P., Gries, D., Luckham, D., Moler, C., Pnueli, A., Seegmüller, G., Stoer, J., Wirth, N., Günther, C.G. (eds.) EUROCRYPT 1988. LNCS, vol. 330, pp. 275–280. Springer, Heidelberg (1988). https://doi.org/10.1007/3-540-45961-8_25
18. McEliece, R.J.: A public-key cryptosystem based on algebraic coding theory. DSN Progress Report, pp. 114–116 (1978)
19. Misoczki, R., Tillich, J.P., Sendrier, N., Barreto, P.S.L.M.: MDPC-McEliece: new McEliece variants from moderate density parity-check codes. In: 2013 IEEE International Symposium on Information Theory (ISIT), pp. 2069–2073, July 2013
20. Niederreiter, H.: Knapsack-type cryptosystems and algebraic coding theory. Problems Control Inf. Theory **15**, 159–166 (1986)
21. Nilsson, A., Johansson, T., Stankovski, P.: Error amplification in code-based cryptography. IACR Trans. Cryptogr. Hardw. Embed. Syst. **2019**(1), 238–258 (2018)
22. Paiva, T., Terada, R.: Improving the efficiency of a reaction attack on the QC-MDPC McEliece. IEICE Trans. Fundam. Electron. Commun. Comput. Sci. **E101.A**, 1676–1686 (2018)
23. Prange, E.: The use of information sets in decoding cyclic codes. IRE Trans. Inf. Theory **8**(5), 5–9 (1962)
24. Santini, P., Baldi, M., Cancellieri, G., Chiaraluce, F.: Hindering reaction attacks by using monomial codes in the McEliece cryptosystem. In: 2018 IEEE International Symposium on Information Theory (ISIT), pp. 951–955, June 2018
25. Santini, P., Baldi, M., Chiaraluce, F.: Assessing and countering reaction attacks against post-quantum public-key cryptosystems based on QC-LDPC codes. In: Camenisch, J., Papadimitratos, P. (eds.) CANS 2018. LNCS, vol. 11124, pp. 323–343. Springer, Cham (2018). https://doi.org/10.1007/978-3-030-00434-7_16
26. Stern, J.: A method for finding codewords of small weight. In: Cohen, G., Wolfmann, J. (eds.) Coding Theory 1988. LNCS, vol. 388, pp. 106–113. Springer, Heidelberg (1989). https://doi.org/10.1007/BFb0019850
27. Tillich, J.P.: The decoding failure probability of MDPC codes. In: 2018 IEEE International Symposium on Information Theory (ISIT), pp. 941–945, June 2018

On IND-CCA1 Security of Randomized McEliece Encryption in the Standard Model

Franz Aguirre Farro$^{(\boxtimes)}$ and Kirill Morozov

Department of Computer Science and Engineering, University of North Texas,
Denton, TX 76207, USA
FranzAguirreFarro@my.unt.edu, Kirill.Morozov@unt.edu

Abstract. We show that a modification of Nojima et al.'s Randomized McEliece Encryption (RME), where receiver verifies the correctness of the decrypted plaintext, achieves IND-CCA1 security in the standard model. We rely on the two (standard) assumptions used also for RME: hardness of general decoding and Goppa code indistinguishability (sometimes they are jointly referred to as "the McEliece assumptions"), plus an extra assumption on non-falsifiability of the McEliece ciphertexts. The later one implies that an adversary is unable to sample a McEliece ciphertext for which the message and error vector are unknown. This assumption is non-standard, however it represents a win-win argument, in the sense that breaking it would imply efficient sampling of McEliece ciphertexts, which in turn may potentially lead us to a Full Domain Hash code-based signature based on the McEliece PKE—without rejection sampling as in the Courtois-Finiasz-Sendrier signature from Asiacrypt 2001—a long-standing open problem in code-based cryptography.

Keywords: McEliece public-key encryption ·
Non-adaptive chosen-ciphertext attacks · IND-CCA1 · Standard model

1 Introduction

The importance of code-based public-key encryption (PKE) is evidenced by strong candidates in the post-quantum standardization process organized by NIST [26]. The McEliece PKE [19] introduced in 1978 was the first cryptosystem based on error-correction codes. Security of this scheme has been extensively studied in the past years—see, e.g., the surveys [9,21,28].

In particular, Kobara and Imai [15] considered adaptations of the Fujisaki-Okamoto transform [12] to the McEliece PKE that achieve security against adaptive chosen-ciphertext attacks (IND-CCA2), in the random oracle (RO) model [1]. Although RO methodology generally results in efficient schemes, it may be beneficial to avoid it not only because some difficulty may arise with instantiating RO [2], but also because a new direction for quantum attacks focusing on RO is hereby introduced [4]. In this work, we will avoid the use of RO, that is we will focus on "the standard model".

© Springer Nature Switzerland AG 2019
M. Baldi et al. (Eds.): CBC 2019, LNCS 11666, pp. 137–148, 2019.
https://doi.org/10.1007/978-3-030-25922-8_8

1.1 Related Works

In [27], Nojima et al. proved that the McEliece PKE with random padding of a plaintext, called the Randomized McEliece PKE, is secure against chosen-plaintext attacks (IND-CPA) in the standard model. The proof is based on hardness of the Learning Parity with Noise (LPN) and the Goppa code indistinguishability. The later assumption implies that a public key of the McEliece PKE is indistinguishable from random. It is valid with respect to parameters used for the McEliece PKE (the rate of the underlying Goppa code is around 0.5), although it was shown not to hold for the high-rate Goppa codes [10]. We emphasize that security against chosen-ciphertext attacks (CCA) was not considered in [27], and therefore no checks on the ciphertext were made in their construction.

Dowsley et al. [8] presented an IND-CCA2 code-based PKE in the standard model. Their scheme can be seen as combination of the following two approaches applied to the McEliece PKE: the Dolev-Dwork-Naor construction [7] and the Rosen and Segev k-repetition paradigm [31]. Due to employing the McEliece PKE's plaintext-checkable property (discussed below), they avoid the use of the non-interactive zero-knowledge (NIZK) proofs. Still, it may be difficult to use their scheme in practice due to k-repetition. Recently, Persichetti [30] presented a direct application of the Rosen-Segev paradigm to the McEliece PKE, but still the k-repetition results in large public keys and plaintexts.

Mathew et al. [18] presented an efficient IND-CCA2 code-based PKE in the standard model. As a basis for their construction they employed the Niederreiter PKE [25], which can be seen as a dual version [9] of the McEliece PKE. Their scheme uses the approach similar to the all-but-one trapdoor functions by Peikert and Waters [29] to reduce k-repetition to 2-repetition, roughly speaking. Still, the public-key and ciphertext sizes roughly double, as compared to the original Niederreiter PKE, and also a one-time signature is used.

1.2 Our Contribution

To the best of our knowledge, security of the McEliece PKE against non-adaptive chosen ciphertext attacks (IND-CCA1) has not been studied in the current literature. In the IND-CCA1 scenario, an adversary has access to the decryption oracle only before obtaining the challenge ciphertext. This notion of security does not imply non-malleability and one may expect that it can be achieved using a simpler construction as required for the IND-CCA2 security. Still, IND-CCA1 is a meaningful notion as it provides certain security guarantee for the PKE schemes when used, e.g., in the context of electronic voting [16].

Normally, IND-CCA1 security requires a certain redundancy in the ciphertext which can be used by the decryption algorithm to recognize valid ciphertexts. The latter will prevent leakage of the information on the secret key to the adversary who is allowed to query the decryption oracle (before seeing the challenge ciphertext). We show a somewhat surprising result that the McEliece encryption has an inherent redundancy for such recognizing. It can be done using a natural algorithm, which we describe in Sect. 3 as PCheck (from "plaintext check"). It is perhaps a good idea to integrate it into all implementations of the McEliece PKE. The algorithm PCheck takes a decrypted plaintext and checks it against the ciphertext. The Randomized McEliece encryption equipped with PCheck will be called the "Randomized-and-Checked McEliece encryption" (RCME).

Verifying the contents of the RME ciphertext is possible due to the plaintext-checkable property of the McEliece PKE, which seems to be folklore—we informally describe it next. Given an alleged message m', and knowing that the ciphertext must have a form $c = mG^{pub} + e$, where G^{pub} is the public key and e is the error vector of weight t, one can compute a vector $y = m'G^{pub}$, add it bitwise to c, and expect to see the result being of weight exactly t, if m' was indeed the plaintext (see Sect. 3 for details). The PCheck algorithm simply applies the above procedure to the result of decryption. Note that this reasoning applies to the RME scheme as well, if we consider the padding to be a part of the message.

This approach is unconventional because the result of decrypting an arbitrary vector (probably an invalid ciphertext) might be unpredictable. Luckily, we have a property that only one possible message matches[1] a valid McEliece ciphertext [13, 22]. This implies that no matter what the decryption algorithm outputs, it is either a correct plaintext or a vector, which will be certainly rejected by PCheck.

Our proof relies on the same assumptions as the RME scheme [27], plus an extra assumption on non-falsifiability of the McEliece ciphertexts. This implies that an adversary is unable to sample a McEliece ciphertext for which the message and error vector are unknown. Roughly speaking, we assume that the only way for the adversary to generate ciphertexts is to use the McEliece encryption algorithm. This assumption is non-standard but it has been used in the context of IND-CCA1 security before, i.e., in the proof of Damgård's ElGamal scheme [6, 16, 24]. Moreover, this assumption represents a win-win argument in our context, in the sense that breaking it would imply an efficient sampling of the McEliece ciphertexts, which in turn may potentially lead us to an efficient Full Domain Hash (FDH) code-based signature based on the McEliece PKE—without rejection sampling as in the Courtois-Finiasz-Sendrier signature [5]—a long-standing open problem in code-based cryptography.

1.3 Organization

In Sect. 2, we describe known assumptions and useful definitions. Section 3 contains a description of RCME—our variant of the Randomized McEliece encryption equipped with plaintext checks. Also, in this section, we put forward a new code-based assumption which in fact follows directly from previous work but simplifies our proofs, as well as a new non-falsifiability assumption. Section 4 contains the proof of IND-CCA1 security for RCME, and Sect. 5 concludes our work with a discussion of future works and open questions.

---

[1] In the context of commitments, this follows from a perfect binding of the commitment scheme based on the McEliece PKE [22].

2   Preliminaries

In this section, we partly rely on the exposition of [21]. We begin with introducing some useful notation.

A concatenation of vectors $x \in \mathbb{F}_2^{k_0}$ and $y \in \mathbb{F}_2^{k_1}$ is written as $(x|y) \in \mathbb{F}_2^{k_0+k_1}$. For $x, y \in \mathbb{F}_2^n$, $d_H(x,y)$ denotes the Hamming distance between x and y, that is the number of positions where they differ. The Hamming distance of $x \in \mathbb{F}_2^n$ to the zero-vector 0^n, which is denoted by $w_H(x) := d_H(x, 0^n)$, is called the *Hamming weight* of x. We will write "0" to represent the zero-vector 0^n, omitting n that will be clear from the context. For $x, y \in \mathbb{F}_2^n$, $x + y$ denotes a bitwise exclusive-or. Let us define the set of n-bit vectors of Hamming weight t as $\mathcal{E}_t^n$, i.e., $\mathcal{E}_t^n := \{x \in \mathbb{F}_2^n \mid w_H(x) = t\}$. We denote by $x \leftarrow \mathcal{X}$ a uniformly random selection of an element from its domain $\mathcal{X}$.

Probabilistic polynomial time is abbreviated as PPT.

2.1   Linear Codes

A binary (n, k)-code $\mathcal{C}$ is a k-dimensional subspace of the vector space $\mathbb{F}_2^n$; n and k are called the *length* and the *dimension* of the code, respectively. We call $\mathcal{C}$ an (n, k, d)-code, if its so-called *minimum distance* is

$$d := \min_{\substack{x,y \in \mathcal{C} \\ x \neq y}} d_H(x, y).$$

For any codeword x of the code that corrects t errors, it holds [17] that $w_H(x) \geq 2t + 1$. For more information on coding theory, see [17,32].

2.2   Security Assumptions and Definitions

Definition 1 (General Decoding Problem)
Input: $G \leftarrow \mathbb{F}_2^{k \times n}$, $c \leftarrow \mathbb{F}_2^n$ and $0 < t \in \mathbb{N}$.
Decide: If there exists $x \in \mathbb{F}_2^k$ such that $e = xG + c$ and $w_H(e) \leq t$.

This problem was shown to be NP-complete by Berlekamp et al. [3]. No efficient (polynomial-time) algorithm is known for solving it in the average case, when using recommended parameters [9,11]. We assume hardness of general decoding in the average case.

Definition 2 (Goppa Code Distinguishing Problem)
Input: $G \in \mathbb{F}_2^{k \times n}$.
Decide: Is G a parity-check matrix of an (n, k) irreducible Goppa code, or of a random (n, k)-code?

An important step towards solving the above problem was made by Faugère et al. [10] by introducing a distinguisher for the high-rate Goppa codes[2] (i.e.,

---

[2] Such the codes are not typically used for public-key encryption, but rather for constructing code-based digital signatures [5].

when k is close to n), however this distinguisher is not known to work for typical parameters of the McEliece PKE. Nonetheless, it shows that the above problem must be used with extra care. We assume hardness of the Goppa code distinguishing problem for the parameters related to the McEliece encryption [11].

We take the next definition from [14].

Definition 3 (IND-CCA1 security). *A public-key encryption scheme is secure against non-adaptive chosen-ciphertext attacks (IND-CCA1), if for any two-stage PPT adversary $\mathcal{A} = (\mathcal{A}_1, \mathcal{A}_2)$ with security parameter κ, in the following experiment the advantage is negligible:*

$$\begin{aligned}
&\text{Exp}_{\text{PKE},\mathcal{A}}^{CCA1}(\kappa) \\
&(pk, sk) \leftarrow \text{Gen}\,(1^\kappa) \\
&(m_0, m_1, \ state\ ) \leftarrow \mathcal{A}_1^{\text{Dec}(sk,\cdot)}(pk) \quad s.t. \quad |m_0| = |m_1| \\
&b \leftarrow \{0,1\} \\
&c^* \leftarrow \text{Enc}\,(pk, m_b) \\
&b' \leftarrow \mathcal{A}_2\,(c^*, state) \\
&\quad if\ b = b'\ return\ 1,\ else\ return\ 0.
\end{aligned}$$

The attacker wins the game if $b = b'$. We define the advantage of $\mathcal{A}$ in the experiment as:

$$\text{Adv}_{\text{PKE},\mathcal{A}}^{CCA1}(\kappa) = \left| \Pr\left[\text{Exp}_{\text{PKE},\mathcal{A}}^{CCA1}(\kappa) = 1 \right] - \tfrac{1}{2} \right|.$$

3   Strengthening the Randomized McEliece Encryption

We present the Randomized-and-Checked McEliece PKE scheme (abbreviated as RCME)—a simple modification of the RME scheme [27], which features checking of the plaintext, randomness, and error vector obtained as the result of decryption against the ciphertext. First, we describe the encryption scheme, and then the checking algorithm denoted as "PCheck".

Definition 4 (Randomized-and-Checked McEliece PKE scheme). This scheme, which uses system parameters $n, k, t \in \mathbb{N}$ consists of the following triplet of algorithms $(\mathcal{K}, \mathcal{E}, \mathcal{D})$:

- Key generation algorithm $\mathcal{K}$:
  On input n, k, t, generate the following matrices:
  - $G \in \mathbb{F}_2^{k \times n}$ – the generator matrix of an irreducible binary Goppa code correcting up to t errors. Its decoding algorithm is denoted as Dec.
  - $S \in \mathbb{F}_2^{k \times k}$ – a random non-singular matrix.
  - $P \in \mathbb{F}_2^{n \times n}$ – a random permutation matrix (of size n).
  - $G^{pub} = SGP \in \mathbb{F}_2^{k \times n}$.

  Output the public key $pk = (G^{pub}, t)$ and the secret key $sk = (S, G, P, \text{Dec})$.

- Encryption algorithm $\mathcal{E}$: On input a plaintext $m \in \mathbb{F}_2^{k_1}$ and the public key pk, choose a vector $r \leftarrow \mathbb{F}_2^{k_0}$ (to serve as a plaintext padding), where $k = k_0 + k_1$, where the integer parameters k_0 and k_1 are chosen as directed by RME [27], and $e \leftarrow \mathcal{E}_t^n$ to serve as an error vector, and output the ciphertext

$$c = (r|m)G^{pub} + e.$$

- Decryption:
    - Using P, compute $cP^{-1} = (r|m)SG + eP^{-1}$.
    - Run $\mathsf{Dec}(cP^{-1})$.
      If $\mathsf{Dec}(cP^{-1})$ outputs "$\perp$", then output "$\perp$".
    - Solve the linear system represented by $(r|m)SG$ and SG for $(r|m)$.
    - If $\mathsf{PCheck}(G^{pub}, (r|m), c) = 0$ then output "$\perp$", otherwise output m.

We emphasize that it is possible in principle to arrange that the decoding algorithm produces an output in polynomially many steps, no matter what the input was. The reason is that the running time of the decoding algorithms is known (see, e.g., [9]), and therefore one may always output "$\perp$" when an upper bound on the running time is exceeded. It is possible that in each specific case, there may be more efficient ways to ensure that the output (the decoded message or "$\perp$") is always delivered, but this is out of scope of this work.

The algorithm $\mathsf{PCheck}(G^{pub}, v, c)$ is described next. The vector $v \in \mathbb{F}_2^k$ denotes an alleged ciphertext that is contained in the ciphertext c.

$$\mathsf{PCheck}(G^{pub}, v, c)$$
$$c' = vG^{pub} + c$$
$$\text{if } w_H(c') = t, \text{ then output } 1$$
$$\text{else output } 0$$

For completeness, we provide an easy proof of the fact that the above algorithm works correctly. The formula for c' can be written as

$$c' = vG^{pub} + c = ((r|m) + v)G^{pub} + e.$$

If $(r|m) = v$, then $(r|m) + v = 0$ and hence $c' = e$, so we must have $w_H(c') = t$. Otherwise, when $(r|m) \neq v$, the vector $((r|m) + v)G^{pub}$ represents a codeword generated by G^{pub}, and since the Goppa code by construction corrects (up to) t errors, its codeword must be of weight at least $2t + 1$, i.e., we have

$$w_H[((r|m) + v)G^{pub}] \geq 2t + 1.$$

Now, the above implies that

$$w_H[((r|m) + v)G^{pub} + e] \geq t + 1,$$

because the smallest weight is achieved when the positions of 1's in e precisely match those in $((r|m) + v)G^{pub}$, which results in weight $t + 1$. Therefore, PCheck always outputs 0, if v is not the contents of c.

3.1   New Assumptions

The following assumption simplifies our proof, it follows directly from the results of [27]. Let us introduce some notation first. Denote by $G_r^{pub} \in \mathbb{F}_2^{k_0 \times n}$ the submatrix of G^{pub}, which corresponds to the random padding, and by $G_m^{pub} \in \mathbb{F}_2^{k_1 \times n}$ the submatrix of G^{pub}, which corresponds to the plaintext. Hereby, the RME/RCME ciphertext can be written as

$$c = (r|m)G^{pub} + e = rG_r^{pub} + mG_m^{pub} + e = rG_r^{pub} + e + mG_m^{pub}. \qquad (1)$$

Definition 5 (Decisional Randomized McEliece (DRME) Assumption).
In the setting of the Randomized McEliece PKE, $rG_r^{pub} + e$ is indistinguishable from random.

Lemma 1 ([27]). DRME assumption is implied by the LPN and Goppa Indistinguishability Assumptions.

Next, we explain that in the previous lemma, we can replace the LPN problem with the General Decoding problem. This issue is just a technicality, the reason for this replacement is that the General Decoding problem is more commonly used in the code-based cryptography literature.

Let us denote the vector $rG_r^{pub} + e$ appearing in Definition 5 as the DRME vector. We observe that the proof in [27] proceeds by: (1) Replacing G_r^{pub} in the DRME vector with the random matrix, and (2) Replacing the random fixed-weight error vector e with that of the Bernoulli distribution (with appropriate parameters) hereby obtaining the LPN vector. We observe that at Step (1), we have exactly the instance of General Decoding problem as stated in Definition 2, and therefore we have the following:

Lemma 2. DRME assumption is implied by the General Decoding and Goppa Indistinguishability Assumptions.

The proof of the above lemma follows by a straightforward rearrangement of the arguments in [27].

Definition 6 (Non-Falsifiable   Knowledge   of   McEliece   Plaintext (NFKMP) Assumption)
There exists no PPT algorithm $\mathcal{A}$ as follows:
Input: A McEliece public key (G^{pub}, t), where $G^{pub} \in \mathbb{F}_2^{k \times n}$, $0 < t \in \mathbb{N}$;
Output: $c \in \mathbb{F}_2^n$ such that $c = vG^{pub} + e$, where $v \in \mathbb{F}_2^k$, $e \in \mathcal{E}_t^n$.
Except for the case, when there exists a PPT algorithm $\mathcal{B}$, which has access to $\mathcal{A}$ as a sub-routine, and works as follows:
Input: Same as the input and output of $\mathcal{A}$;
Output: $v \in \mathbb{F}_2^k$ and $e \in \mathcal{E}_t^n$, such that $c = vG^{pub} + e$.

In the above definition, we assume hardness of sampling of McEliece ciphertexts for which plaintexts (and hence error vectors) are unknown. Now, suppose that an efficient sampling of this kind was known. Then, this would potentially imply an FDH-like digital signature constructed from the McEliece PKE in the natural way: Map a message to the random coins (using some hash function), then use them to sample the ciphertext, then decrypt it and use the result as a signature. The evidence supporting that the NFKMP assumption is plausible is the fact that such the signature has not been devised in the past 40 years. We emphasize that although the Courtois-Finiasz-Sendrier signature [5] does use the FDH paradigm (with rejection sampling), it requires very specific parameters (i.e., very high-rate codes). This leads to penalties both in efficiency and, at least potentially, in security [10], although distinguishability of the high-rate Goppa codes per se was shown not to be a security issue for this scheme [23]. On the positive side, one may see this as a "win-win" argument in the sense that if this assumption was refuted, it might potentially lead to efficient signatures based on a well-studied cryptosystem.

4   Main Result

Theorem 1. Under the hardness of General Decoding, Goppa Indistinguishability, and NFKMP Assumptions, the Randomized-and-Checked McEliece public-key encryption scheme is secure against non-adaptive chosen-ciphertext attacks (IND-CCA1) in the standard model.

Proof. In this proof, we follow the lines of Katz' lecture notes [14, Sect. 4] where the IND-CCA1 security proof of the Cramer-Shoup-"Lite" cryptosystem was presented.

Assuming a PPT adversary A in the RCME scheme, the distinguisher algorithm $\hat{A}$ tries to distinguish a DRME vector from a random one. If the tuple given to $\hat{A}$ is a random tuple, then A will have no information about the message, since it is encrypted using a one-time pad (with perfect security). For short, we denote the RCME secret key as sk.

$$\hat{A}(\cdot)$$
$$(G^{pub}, sk) \leftarrow Gen(1^{\kappa})$$
$$r \leftarrow \{0,1\}^{k_0}, \ e \leftarrow \mathcal{E}_t^n, \ y = rG_r^{pub} + e$$
$$(m_0, m_1) \leftarrow A^{\mathcal{D}_{sk}}(G^{pub})$$
$$b \leftarrow \{0,1\}$$
$$c = y + m_b G_m^{pub}$$
$$b' \leftarrow A(G^{pub}, c)$$
$$\text{if } b = b', \text{ then guess "DRME vector"}$$
$$\text{else guess "random vector"}$$

Claim. If $\hat{A}$ gets a DRME tuple, then A's view of the game is the same as in an execution of the real RCME encryption scheme.

Proof. We observe that the public-secret key pair generated by the distinguisher $\hat{A}$ is identical to the key pair generated in the real execution of the RCME scheme, and therefore the decryption queries will be answered exactly as in the real execution of RCME. Now, according to (1), the real RCME ciphertext can be written as $c = (r|m)G^{pub} + e = (rG_r^{pub} + e) + mG_m^{pub}$, where the term in brackets is the DRME vector, and hence the challenge ciphertext in the real execution is distributed identically to the ciphertext constructed in the description of $\hat{A}$. $\square$

The above claim implies that

$$\Pr\left[\hat{A} \text{ outputs "DRME tuple"}|\text{DRME tuple}\right] = \Pr\left[b' = b|A \text{ attacks real scheme}\right].$$

Claim. If $\hat{A}$ gets a random vector, then except with negligible probability, A has no information about the bit b chosen by $\hat{A}$, in the information-theoretical sense, assuming that $\hat{A}$ makes polynomially many queries to the decryption oracle.

Proof. The claim follows by observing that $c = y + m_b G_m^{pub}$, where y is a uniformly random vector (chosen independently from m_b) represents a one-time pad encryption scheme, which is perfectly secure.

It remains to show that the decryption queries do not provide additional information to the adversary. Let us call a query "legal" if the corresponding ciphertext is of a correct form, i.e., $c = vG^{pub} + e$, where $v \in \mathbb{F}_2^k$ and $e \in \mathcal{E}_t^n$; and let us call a query "illegal" otherwise.

Observe that the PCheck algorithm described in Sect. 3 leads to rejection of illegal queries with probability 1. Indeed, if the decoding result does not match the ciphertext c, this means that the ciphertext was invalid. This follows from the fact that there exists a unique pair of vectors (r, m) such that $c = (r|m)G^{pub} + e$ holds with $e \in \mathcal{E}_t^n$. This is a simple consequence of the perfect binding of the commitment scheme based on McEliece PKE [22]. This shows that only legal queries consisting of the valid ciphertexts will be replied. Now, the NFKMP assumption (Definition 6) implies that $\hat{A}$ must know the vectors (r, m, e) when making a query, and hence making legal queries does not provide any additional information to the adversary.

The above claim implies that the probability that A correctly guesses b is negligibly close to $1/2$ and therefore $\Pr\left[\hat{A} \text{ outputs "DRME tuple"}| \text{ random tuple}\right]$ is negligibly close to $1/2$ as well. Thus, the advantage of $\hat{A}$ is negligibly close to:

$$\left|\Pr\left[b = b'|A \text{ attacks real scheme }\right] - 1/2\right|.$$

Since the DRME assumption implies that the advantage of $\hat{A}$ is negligible, it follows that the probability that A correctly guesses the value of b when attacking the real scheme is negligibly close to $1/2$, which is the definition of IND-CCA1 security (Definition 3). $\square$

5   Concluding Remarks

It would be interesting to extend our result to IND-CCA2 security in the standard model by adding some extra checks, that would avoid the use of the DDN-like construction as in Dottling et al.'s scheme [8], and avoid the use of complex public keys as in Mathew et al.'s scheme [18].

Another interesting question is to prove that our assumptions are equivalent to assuming the RMCE scheme to be IND-CCA1 (i.e. proving the converse to our main theorem) as it was shown by Lipmaa [16] for Damgård's ElGamal scheme [6].

Finally, one may observe that our construction is generic enough to potentially encompass other variants of the McEliece-type encryption, for example, cryptosystems based on MDPC codes [20]. Generalization of our result to these schemes is one direction for possible future work.

Acknowledgements. We would like to thank the anonymous reviewers for their helpful comments.

References

1. Bellare, M., Rogaway, P.: Random oracles are practical: a paradigm for designing efficient protocols. In: ACM Conference on Computer and Communications Security 1993, pp. 62–73, ACM (1993)
2. Bellare, M., Boldyreva, A., Palacio, A.: An uninstantiable random-oracle-model scheme for a hybrid-encryption problem. In: Cachin, C., Camenisch, J.L. (eds.) EUROCRYPT 2004. LNCS, vol. 3027, pp. 171–188. Springer, Heidelberg (2004). https://doi.org/10.1007/978-3-540-24676-3_11
3. Berlekamp, E., McEliece, R., van Tilborg, H.: On the inherent intractability of certain coding problems. IEEE Trans. Inf. Theory **24**, 384–386 (1978)
4. Boneh, D., Dagdelen, Ö., Fischlin, M., Lehmann, A., Schaffner, C., Zhandry, M.: Random oracles in a quantum world. In: Lee, D.H., Wang, X. (eds.) ASIACRYPT 2011. LNCS, vol. 7073, pp. 41–69. Springer, Heidelberg (2011). https://doi.org/10.1007/978-3-642-25385-0_3
5. Courtois, N.T., Finiasz, M., Sendrier, N.: How to achieve a McEliece-based digital signature scheme. In: Boyd, C. (ed.) ASIACRYPT 2001. LNCS, vol. 2248, pp. 157–174. Springer, Heidelberg (2001). https://doi.org/10.1007/3-540-45682-1_10
6. Damgård, I.: Towards practical public key systems secure against chosen ciphertext attacks. In: Feigenbaum, J. (ed.) CRYPTO 1991. LNCS, vol. 576, pp. 445–456. Springer, Heidelberg (1992). https://doi.org/10.1007/3-540-46766-1_36
7. Dolev, D., Dwork, C., Naor, M.: Nonmalleable cryptography. SIAM J. Comput. **30**(2), 391–437 (2000)
8. Döttling, N., Dowsley, R., Müller-Quade, J., Nascimento, A.C.A.: A CCA2 secure variant of the McEliece cryptosystem. IEEE Trans. Inf. Theory **58**(10), 6672–6680 (2012)
9. Engelbert, D., Overbeck, R., Schmidt, A.: A summary of McEliece-type cryptosystems and their security. J. Math. Cryptol. **1**, 151–199 (2007)
10. Faugère, J., Gauthier-Umaña, A., Otmani, V., Perret, L., Tillich, J.: A distinguisher for high rate McEliece cryptosystems. In: Information Theory Workshop 2011, pp. 282–286. IEEE (2011)
11. Finiasz, M., Sendrier, N.: Security bounds for the design of code-based cryptosystems. In: Matsui, M. (ed.) ASIACRYPT 2009. LNCS, vol. 5912, pp. 88–105. Springer, Heidelberg (2009). https://doi.org/10.1007/978-3-642-10366-7_6
12. Fujisaki, E., Okamoto, T.: Secure integration of asymmetric and symmetric encryption schemes. In: Wiener, M. (ed.) CRYPTO 1999. LNCS, vol. 1666, pp. 537–554. Springer, Heidelberg (1999). https://doi.org/10.1007/3-540-48405-1_34

13. Jain, A., Krenn, S., Pietrzak, K., Tentes, A.: Commitments and efficient zero-knowledge proofs from learning parity with noise. In: Wang, X., Sako, K. (eds.) ASIACRYPT 2012. LNCS, vol. 7658, pp. 663–680. Springer, Heidelberg (2012). https://doi.org/10.1007/978-3-642-34961-4_40

14. Katz, J.: Lecture Notes on Advanced Topics in Cryptography (CMSC 858K), Lecture 9, 24 February 2004

15. Kobara, K., Imai, H.: Semantically secure McEliece public-key cryptosystems - Conversions for McEliece PKC-. In: Kim, K. (ed.) PKC 2001. LNCS, vol. 1992, pp. 19–35. Springer, Heidelberg (2001). https://doi.org/10.1007/3-540-44586-2_2

16. Lipmaa, H.: On the CCA1-Security of Elgamal and Damgård's Elgamal. In: Inscrypt 2010, pp. 18–35 (2010)

17. MacWilliams, F., Sloane, N.J.A.: The Theory of Error-Correcting Codes. North-Holland, Amsterdam (1992)

18. Preetha Mathew, K., Vasant, S., Venkatesan, S., Pandu Rangan, C.: An efficient IND-CCA2 secure variant of the Niederreiter encryption scheme in the standard model. In: Susilo, W., Mu, Y., Seberry, J. (eds.) ACISP 2012. LNCS, vol. 7372, pp. 166–179. Springer, Heidelberg (2012). https://doi.org/10.1007/978-3-642-31448-3_13

19. McEliece, R.J.: A public-key cryptosystem based on algebraic coding theory. Deep Space Network Progress Report (1978)

20. Misoczki, R., Tillich, J.-P., Sendrier, N., Barreto, P.S.L.M.: MDPC-McEliece: new McEliece variants from moderate density parity-check codes. ISIT 2013: 2069–2073 (2013)

21. Morozov, K.: Code-based public-key encryption. In: Nishii, R., Ei, S., Koiso, M., Ochiai, H., Okada, K., Saito, S., Shirai, T. (eds.) A Mathematical Approach to Research Problems of Science and Technology. MI, vol. 5, pp. 47–55. Springer, Tokyo (2014). https://doi.org/10.1007/978-4-431-55060-0_4

22. Morozov, K., Roy P.S., Sakurai, K.: On unconditionally binding code-based commitment schemes. In: IMCOM 2017, vol. 101 (2017)

23. Morozov, K., Roy, P.S., Steinwandt, R., Xu, R.: On the security of the Courtois-Finiasz-Sendrier signature. Open Math. **16**(1), 161–167 (2018)

24. Naor, M.: On cryptographic assumptions and challenges. In: Boneh, D. (ed.) CRYPTO 2003. LNCS, vol. 2729, pp. 96–109. Springer, Heidelberg (2003). https://doi.org/10.1007/978-3-540-45146-4_6

25. Niederreiter, H.: Knapsack-type cryptosystems and algebraic coding theory. Prob. Control Inf. Theory **15**(2), 159–166 (1986)

26. NIST Post-Quantum Cryptography Standardization. Round 2 Submissions. 31 January 2019. https://csrc.nist.gov/projects/post-quantum-cryptography/round-2-submissions

27. Nojima, R., Imai, H., Kobara, K., Morozov, K.: Semantic security for the McEliece cryptosystem without random oracles. Des. Codes Crypt. **49**(1–3), 289–305 (2008)

28. Overbeck, R., Sendrier, N.: Code-based cryptography. In: Bernstein, D.J., Buchmann, J., Dahmen, E. (eds.) Post-Quantum Cryptography, pp. 95–145. Springer, Berlin (2009). https://doi.org/10.1007/978-3-540-88702-7_4

29. Peikert, C., Waters, B.: Lossy trapdoor functions and their applications. In: STOC, pp. 187–196 (2008)

30. Persichetti, E.: On the CCA2 security of McEliece in the standard model. In: Baek, J., Susilo, W., Kim, J. (eds.) ProvSec 2018. LNCS, vol. 11192, pp. 165–181. Springer, Cham (2018). https://doi.org/10.1007/978-3-030-01446-9_10

31. Rosen, A., Segev, G.: Chosen-ciphertext security via correlated products. In: Reingold, O. (ed.) TCC 2009. LNCS, vol. 5444, pp. 419–436. Springer, Heidelberg (2009). https://doi.org/10.1007/978-3-642-00457-5_25
32. Roth, R.: Introduction to Coding Theory. Cambridge University Press, Cambridge (2006)

Author Index

Printed in the United States
By Bookmasters